MIND, LIFE, AND UNIVERSE

Conversations with Great Scientists of Our Time

MIND, LIFE, AND UNIVERSE

Conversations with Great Scientists of Our Time

EDITED BY
Lynn Margulis *and* Eduardo Punset

CHELSEA GREEN PUBLISHING COMPANY
WHITE RIVER JUNCTION, VERMONT

A Sciencewriters Book

scientific knowledge through enchantment
Sciencewriters Books is an imprint of Chelsea Green Publishing. Founded and codirected by Lynn Margulis and Dorion Sagan, Sciencewriters is an educational partnership devoted to advancing science through enchantment in the form of the finest possible books, videos, and other media.

Editor-in-chief: John Barstow
Production Director: Bill Bokermann
Project Editor: Collette Leonard
Developmental Editors: Judith Herrick Beard, Dianne Bilyak
Copy Editor: Cannon Labrie
Proofreader: Robin Catalano
Indexer: Marc Schaefer
Designer: Peter Holm, Sterling Hill Productions

Printed in the United States of America
First printing, July 2007
10 9 8 7 6 5 4 3 2 1 07 08 09 10

Our Commitment to Green Publishing
Chelsea Green sees publishing as a tool for cultural change and ecological stewardship. We strive to align our book manufacturing practices with our editorial mission, and to reduce the impact of our business enterprise on the environment. We print our books and catalogs on chlorine-free recycled paper, using soy-based inks, whenever possible. Chelsea Green is a member of the Green Press Initiative (www.greenpressinitiative.org), a nonprofit coalition of publishers, manufacturers, and authors working to protect the world's endangered forests and conserve natural resources.

Mind, Life, and Universe was printed on Nature's Cream, a 50-percent postconsumer-waste recycled, old-growth-forest-free paper supplied by Maple-Vail.

Library of Congress Cataloging-in-Publication Data
Mind, life, and universe : conversations with great scientists of our time / edited by Lynn Margulis and Eduardo Punset.
p. cm.
Includes index.
ISBN 978-1-933392-61-5 (hardcover) — ISBN 978-1-933392-43-1 (pbk.)
1. Scientists—Interviews. 2. Science. I. Margulis, Lynn, 1938– II. Punset, Eduard, 1936– III. Title.

Q141.M5 2007
509.2'2—dc22

2007018552

Chelsea Green Publishing Company
P.O. Box 428
White River Junction, VT 05001
(802) 295-6300
www.chelseagreen.com

CONTENTS

Part Three: Life on an Animate Planet

Part Four: Toward the Invisible

FOREWORD

Science, the Hidden Force on Society

DAVID T. SUZUKI

We live in a time of astonishing contradiction. Although the average person has access to more information than anyone has ever had in all of history, we seem unable to cope with enormous problems that confront us in an informed, thoughtful way. We fail to recognize that the most powerful force shaping our lives and society today is scientific knowledge as it is applied by industry, medicine, and the military. The world of my childhood has long since disappeared under the impact of polio vaccines, extinction of smallpox, oral contraception, computers, plastics, jet airplane travel, satellites, xerography, weapons of mass destruction, and a long list of other innovations. Scientific insights lead to applications that can irrevocably alter our lives and society and challenge the very definition of what it is to be human. Yet if we were to judge by the amount of space and time devoted to various subjects in newspapers, television, and radio, we would conclude that politics, business, sports, and celebrity are the most important matters in the universe and in our lives. Think of some of the issues of our time that require action: climate change, species extinction, space exploration, endocrine-system disruption, cloning, stem cell research, nanotechnology, marine-fisheries depletion, intelligent machines, and genetically modified organisms (GMOs). Too often, those who have the power to act are ill equipped to assess the technological and scientific information they need. They cannot make optimal decisions for the rest of us. Public discourse about what science is and what scientists do has never been more needed. Yet many, including some who call themselves scientists, are ignorant about how scientific insights, with their strengths, weaknesses, and severe limitations, differ with respect to evidence and verifiability from other forms of knowledge.

The thirty-six interviews in this book give readers insights into the minds of some of the leading scientists in the world and reveal how they think and how they view their own work. As a scientist who has spent more than four decades working in science as well as on peace, the environment, and social justice, and as a journalist in both print and electronic media, mainly television, I have long believed that scientific literacy must be a fundamental goal of education. Here I share a few thoughts on the interface between the public and the scientific community.

In my last year of college in the fall of 1957 we were electrified by the announcement that the Soviet Union had successfully launched into outer space a basketball-sized satellite—Sputnik. I hadn't even known such a Soviet space exploration program existed! In the agonizing months that followed, the United States failed to duplicate the feat as rockets that carried grapefruit-sized satellites blew up on the launchpad or exploded shortly after liftoff. The Russians continued to rack up one "first" achievement in outer space after another: first animal (the dog, Laika); first man (Yuri Gagarin); first team of cosmonauts; first woman cosmonaut (Valentina Tereshkova). Thereby they demonstrated that their country was more advanced than ours in science, engineering, mathematics, and space medicine. The United States responded to the Soviet challenge, pouring millions of dollars into universities for their advanced students and faculty to support research and training to catch up.

It was a golden era for budding scientists like me. Even though I was a Canadian studying in the United States, widely available grants allowed me to continue my studies. We reveled in science and the joys of discovery. I had a talent in an arcane area of fruit fly genetics called "chromosome mechanics." We were proud of pursuing knowledge for its own sake, of pushing back the frontiers of our ignorance as we probed the very edge of human thought. We were taught, and I believed, that science is the most powerful way of knowing, that through science we extend our understanding of how nature works. This knowledge, we believed, was the basis for control and management of the world around us, and would ultimately benefit all of humankind. But as science exploded in the 1960s and 1970s, life did not improve for most people on Earth. We became aware that technologies (nuclear, DDT, CFCs) had unintended effects on the health of humans, other life forms, and the environment.

As a newly minted assistant professor of genetics, I was poised to make my reputation as a researcher when Rachel Carson published *Silent Spring* in 1962. Her seminal book outlined the unexpected consequences of pesticides. As I read the book, it was as if she were speaking to me as a scientist. The message I heard was: "You scientists are clever. You can create powerful compounds like DDT and show that it kills insects without harming plants. But you forget that the lab, a test tube, or a growth chamber is not the real world. In the real world everything is connected to everything else, so when DDT is sprayed onto fields to kill pests, it persists and accumulates in fish, bird, and even human tissues and harms them." I had assumed that the lab was a miniature replica of the world, so that by the study of life, its environment, and anything else under its controlled conditions we could extrapolate our results to the whole world. I then realized that by focus, measurement, and control of only a small part of nature in the laboratory, we acquire profound insights into that fragment of nature. But only that isolated fragment. Carefully controlled laboratory experiments are in the best reductionist tradition. But the isolation of a part from the whole means we remove it from the context within which it exists and interacts. We no longer see the rhythms, patterns, and cycles that impinge on whatever it is in nature and on which it might impinge. We create an "artifact," something that may superficially resemble the rest of the world but can only be a crude approximation. The results of similar manipulation in nature cannot be directly inferred from laboratory observation alone. Sadly, most geneticists and biotechnologists still have not learned this, Rachel Carson's fundamental lesson.

Canada also began to increase research funding as student enrollment in universities rose rapidly in the 1970s. Most new faculty appointments were assured of small research grants, which then gradually increased with time. But many of us argued that to attract and keep good scientists in Canada, our funding had to be competitive with countries like the United States. As government funding to science increased, so came the demand for accountability: For what, exactly, was the money being spent? To what problems were the expected scientific results relevant? In Parliament, Senator Maurice Lamontagne completed his study of the role of research in Canada, and coined the term "mission-oriented" work as he recommended a closer relationship between scientific research and its application. Here began an unexpected consequence of support for the sciences!

Grant applications required statements about what practical "payoffs" might result from the work described in the proposals. In attempts to satisfy the political demand, university faculty scientists perpetrated a false impression of how science works. Thus, someone interested in the rectal temperature of penguins (I make up this example, of course) might suggest his study could reveal a relationship between penguin body temperature and ambient temperature that would permit assessment of the consequences of global warming. We went from proposing experiments to expectations of their results that would lead to "further progress," nationally or in our specialized fields. We suggested, inadvertently and under pressure, that scientific research proceeds in a linear fashion from experiment A to B to C to a cure for cancer or a saleable product. In playing this game of justifying financial support, many scientists have come to believe that research actually proceeds in a linear fashion, so more money is all that is needed to solve most of the world's problems. Nothing could be farther from the truth.

We carry out experiment A in the first place because we want to find out what the results will tell us about nature. We have little idea about the outcome. If science proceeded in a linear fashion toward application there would be no mystery to impel us to become scientists. Life is far less linear: two scientists who work in different areas meet at a pub or party and they exchange ideas. They chat, tease, drink, gossip, and soon discover that their mutual interests and capabilities might be combined to study a familiar preoccupation. Barbara McClintock (1902–92), the great corn geneticist at Cold Spring Harbor Laboratory in New York, studied heredity in hybrid corn. She documented "jumping genes" in the kernels, the food-rich endosperm of maize. Most of us thought that she had seen an interesting quirk of the plant. None of us ever dreamed that her esoteric studies would lead to the creation of tools to study development in hundreds of different animals, plants, and microbes or that she would earn a Nobel Prize in Medicine (1983). In the same way, no one could have anticipated that Werner Arber's elegant analysis of the fascinating genetic puzzle of how bacteria resist infection by viruses would lead to the discovery of restriction enzymes that have made genetic engineering possible and would earn him the Nobel Prize (1978). These were outstanding scientists studying intriguing phenomena in nature, but if their grants had been determined by the demand for relevance, they would never have been funded. Even today, few realize that microbial community analysis such as the study of predatory bacteria in lakes (see chapter 29 with Ricardo Guerrero) and symbiotic bacterial organs in worms would lead to simple cures for ulcers and river blindness. Only scientists who intimately know their objects of study and the relationships among them (whether life forms, analytical equipment, visual images, mathematics equations, or other) are familiar enough with the complexities to debunk the common, woefully inadequate myth that there is a linear relationship between science and application.

> Dr. Faustus, Dr. Frankenstein, Dr. Moreau, Dr. Jekyll, Dr. Cyclops, Dr. Caligari, Dr. Strangelove. The scientist who does not face up to the warning in this persistent folklore of mad doctors is himself the worst enemy of science. In these images of our popular culture resides a legit-

> imate public fear of the scientist's stripped down, depersonalized conception of knowledge—a fear that our scientists, well-intentioned and decent men and women all, will go on being titans who create monsters.

The widespread popular image of the cold, calculating scientist lacking passion and examining the world with detachment is an error. Not only are people suspicious of others they perceive as cold, logical, and devoid of emotion, but scientists, especially those interviewed in this book, are quite the opposite. They are warm, passionate, creative, and thoughtful.

When I was an active *Drosophila* geneticist, janitors often came through the lab at night. Some asked me what I was doing at work so late. When I showed them a fruit fly under the microscope, without exception, they reacted with amazement and wonder at the vivid scarlet color of the eyes. The architectural regularity of the ommatidia, the beautiful and regularly arranged segments in the compound insect eye, dazzled anyone who beheld it. They marveled at the precise placement of bristles and hairs behind the fruit fly's head and on its thorax. This sense of wonder that children share with some janitors is what attracted most of us to science in the first place. Yet if I were to begin one of my scientific articles, "The red-patterned compound eye of a fruit fly is a joy to behold. The brilliant ommatidia are organized with such exquisite mathematical precision and overlaid with such a fine precise pattern of bristles and hairs that it elevates my spirit to study these paired organs," I am certain that the paper would be rejected out-of-hand as too subjective and emotional, and irrelevant to the research. Scientific papers written in a way that fails to expunge the very reason we become scientists and that make science such a wonderful venture are unacceptable to fellow professionals. Our culture has made science and scientists seem to be what they are not at all—emotionless, robotic caricatures of science-fiction movies. This set of interviews is the finest antidote to the profoundly ignorant belief that scientists are robotic animated mannequins. You will see how Eduardo Punset's interviews display the humanity of these remarkable investigators without sacrifice of their authentic and intellectual achievements.

With the enormous commitment of public funds for research, the pressure to come up with applications that will solve problems such as global warming, species extinction, or toxic pollution has increased enormously. So these days, public relations offices, the machines of universities, government labs, and corporations trumpet incremental gains in knowledge as "breakthroughs."

While few discoveries lead directly to applications, the potential return on the investment based on a linear model of scientific discovery is easy to see. But consider this: I graduated with a Ph.D. in genetics in 1961. We knew about the double helix of DNA, about operons and messenger RNA. It was an exciting time to be a geneticist. But when I explain to today's students what the "hottest" new ideas of 1961 were, they burst out laughing. Of course, in 2007, the concepts of 1961 seem in retrospect ridiculous. But those same students are stunned when I show them that their current notions of the truths of molecular genetics will be just as incorrect, indeed ludicrous, twenty years from now. They don't understand that in any revolutionary area, most of our current ideas are wrong. That does not denigrate science; rather it is the true way that science informs us of the world. In any new territory, we make observations and measurements that at first we don't understand. We construct hypotheses that "make sense" of disparate pieces of information. Often we find the same hypotheses in the old scientific literature made by prescient predecessors on much less information. We try to formulate the current version of any idea so specifically that we can test it. When the results are gathered, chances are that we will be forced to abandon our hypothesis and create another. Or maybe we will have to modify it to test it further. We usually prove that any current idea, even of our teachers or our fellow "experts" is too simple as we probe deeper into the mysteries of mind, life, and universe. Scientists should admit out loud more frequently the tentative nature of any current idea. Punset's "experts" recognize their own limitations. Most have a wonderful sense of humor. The beauty of Punset's interviewees is exactly that: they are humble in the face of the vast unknown. They anticipate the inevitable backlash when the public impugns them for failure to deliver neat expected solutions—in return for financial support.

These interviewees delightfully retreat from the "hype" that generates public disappointment. In this collection of thirty-six interviews with thirty-seven scientists, Punset evokes the enthusiasm, insight, obsession, and "what-if's" so that we readers are jolted. We gain deeper insights through their spontaneous responses and opinions in their own words. Leading scientists here tell about their observable world. They are not speaking to any professional peers or colleagues who share years of experience with their own specialty. They speak to this friend, this Spanish television hero, in his nearby armchair. All try hard to explain their difficult esoteric quest. They each succeed with remarkable clarity, largely because of Punset's technique: He asks forthrightly and bluntly questions whose answers interest him. They know that their answers, derived from years of excellent science, provide only tiny windows into a world of which we readers are enormously ignorant. These scientists are very social primates. They collaborate to find ways to apply their intrinsically limited knowledge. They all know that the myth of "more funds," that anyone can generate an immediate solution to a specific problem of the environment or human health simply by throwing money at it, is not the way scientific knowledge expands.

The greatest strength of science is in its "accurate-as-possible description" of a tiny part of nature's great secrets. As we push back the curtains of our preconceptions and lack of understanding we discover all manner of things because we know so little. I was impressed to read about the ingenious ways scientists have teased out insights into primate behavior and human sexual development, the Earth's past temperatures and atmospheres, the evolution of life over millions of years. Punset's scientists don't pontificate; they don't prescribe. They realize that the solutions to problems such as cancer, rapid climate change, or control of the growth of the human population are not even scientific. Rather, such complex social, economic, and political issues transcend the competence of any scientist. Most important, for Punset's interviewees the journey to find solutions to scientific problems is fun, which is why, to begin with, I, too, went into scientific research!

INTRODUCTION AND ACKNOWLEDGMENTS

Lynn Margulis

My claim, one I heard Stewart Brand* say, is that
"science is the only news; all else is hearsay and gossip."

Science as open inquiry and careful analysis of the real properties of the real world seems to us to still be the most powerful, honest, and collectively useful path to knowledge. World-class science, in principle, is available to all people willing to observe and study regardless of social class, race, age, gender, nationality, or other divisive influence. Genuine scientific activity as a way of knowing stands in stark contrast to knowledge via the teachings of mystical or religious beliefs. The intrinsically limited results of worldwide primary science, that done and described by the scientist, inform our opinion, help us enjoy and control our lives, and of course lead to technological advance. But science, an intensely social activity based on free inquiry, is not equivalent to application of science as technology. The core attributes of science include observation and measurement of selected aspects of the physical and living world, honest and open report of the observations and measurements by publication, analysis and criticism of the published results by other scientists, and verification that new results are consistent with previous hard-won bodies of scientific literature and established fact. Our book of conversations with great scientists at the height of their careers is predicated on an assumption: readers want scientific truths much more than they want gossip about the private lives of famous people. We have sought accessible accounts from the researchers themselves on important questions. We do not want half-truths. Science is not science at all until the results of inquiry are written in professional journals. A major problem is that only specialists can read that literature—and then, only in their own fields. How are the rest of us to learn?

Our surmise is that our readers would rather learn about an unpleasant well-established scientific fact than a questionable claim of the cancer cure or a human skeleton on Mars. They do not want hasty accounts from confused journalists about painless surgery without anesthesia. Nor have any of us the

*"Whole Earth Catalog" and "Whole Earth Review" and Point Foundation fame, author of *Media Lab* and the first publisher of "The Gaia hypothesis" by J. E. Lovelock and me (Margulis, L. and J. E. Lovelock. "The atmosphere as circulatory system of the biosphere—The Gaia hypothesis." *CoEvolution Quarterly* 6: 30–40).

drive and patience to study to make opaque, metaphorical scientific jargon translucent. Most of us lack the ability to command the peculiar language, mathematics, esoteric mechanical or inventive skill needed for any scientific task. All of us lack education and expertise in the broad range of fields represented here.

We, the literate public, prefer to learn from authentic scientific investigators. Why? Because we can trust them as special people even though we know them to be special in only one sense. All scientists by definition are specialists; they could not function were they not. But they are special only in the sense that they can be trusted to know a limited set of reliably known nuanced truths. They know such subtle truths but only within their fields of purview. If they are ignorant of these they can't function as investigators, as researchers.

The contributors chosen here are special in an additional sense. Not only does each participate in the generation of new scientific knowledge, but also each is capable of contextual explanation of reliably known ideas in his or her area of genuine expertise. Punset's interviewees, cautious not to promise information they don't have or "facts" they don't really understand, are superb communicators. They are competent to give readers what we all seek: straightforward, comprehensive, comprehensible, but admittedly limited, answers to important questions. Punset determined the importance of the questions (See "The Editors," page 333).

The problem is this: no scientist can give any reader exactly what he or she wants. The scientist, including the remarkable, talented contributors here, can speak with authority about only a minuscule part of the whole. We know so little of the natural world. Our thesis, as we begin face-to-face conversations with selected scientific spokespeople, is that a single person, no matter how talented and knowledgeable, can only brief a reader on a small fraction of that personally familiar to him or her. Therefore we editors try to help. We introduce each scientist's background and face (See "The Scientists" section that begins on page 319). We tie together the thirty-six short interviews by placing them into nine chapters of related themes. Our goal is to distinguish common myth from genuine, well-established scientific truth and to transmit it accurately but personably. Perhaps most important is that we point out the level of confidence in knowledge represented in the various sciences here mainly by our choice of interviewee. Many indicate how the knowledge they seek from the vast unknown might be garnered for a particular case. The detailed methods vary.

Nature guards her secrets jealously. All scholars, when they venture even slightly afield, tend to stumble. Our job is to introduce each theme in context, and to help illuminate the level of confidence in the statements. When we are able we cross-reference the essays or comment on them. In our very short bibliography, we restricted each participant to the choice of no more than two references, usually books. The reading list begins on page 341. We have forced him or her to select the one best, most comprehensive and accessible publication that, in principle, expands the details of background and concepts mentioned in the chapters.

Thus this book is no set of unrelated charming chapters by famous special guests on TV. It is not a list of lofty pronouncements from didactic, incomprehensible experts. Rather the limited nature of authentic scientific inquiry is revealed by each passionate contributor in the context of the big question she asks herself: What is the nature of the universe? What is life? How did life originate and evolve, and what, really, is death? How well does the human mind perceive the mind's physical and social surroundings? The reader can't help but sense the interviewee's idiosyncratic commitment to his own honest inquiry. The eagerness to explain his small piece of the puzzle comes directly from enthusiastic answers. Our brief commentary, bibliographic suggestions, occasional notes, and glossary entries are meant to help readers find their way. The potential seamlessness of the results of scientific inquiry shows, at least in principle, how research might eventually be assembled into a coherent single whole. Science cannot give us easy, universal answers, but research can lead to a truthful approach to the answers of the big questions for all people everywhere. We editors, inspired by the pleasure of our task, integrate by cross-reference conversations that may connect. We attempt to indicate relationships of the scientists' answers to concerns and aspirations of other scientists, readers, and the public. These conversations, like the results of all true scientific inquiry, are open to the public at all times. We invite you to listen in and participate in any way you wish.

The short quotes as epigraphs at the start of many of the interviews are drawn from the speaker who is being interviewed. These thirty-seven thinkers, bluntly queried, tell Eduardo what they really think about the big questions he asks. All the questions have preoccupied each through his or her long career. Often the TV interviews were too long, irrelevancies intervened such that Punset, Margulis, or the translation team omitted portions of the tape. We strove to cut redundancy, incoherent pronouns, and other inevitable distractions of the oral word. In short, verbatim text for television was deliberately edited for comprehensibility. Most of the chapters and comments derive from Eduardo's cache of videotaped interviews and transcripts. Much work by many, especially Elsa Punset (Eduardo's daughter) and her helpers Begonia Barrabés and the editors at Smartplanet, Judith Herrick Beard (of Typro), and Dianne Bilyak (of Greatedit)—as well as John Barstow, Jonathan Teller-Elsberg, Emily Foote, and Collette Leonard at Chelsea Green, helped transform the sloppiness of speech into the flow of modulated, disciplined written form. But the frank and irreverent, humble and funny, distinct approaches, the originality and fresh quips of conversational ripostes were not removed. The internationally dispersed investigators describe their own sciences as they see them in answer to Eduardo, each in his or her own style. We only shortened and tightened the conversations. We did not sanitize them. We replaced the plethora of pronouns ("it," "that," "these," "those who") with their intended referents ("melting glacier," "telescope," "arrogant bureaucrat," "ignorant politician"). We tried not to take liberties but to stick closely to the often-flippant authority of the interviewee.

Everyone in Spain knows that Punset's lively television program *Redes* (*Networks*) charms the TV audience and readers in that country. One cannot walk the streets in the Barrio Gótico, on the Ramblas, or in the Gracia quarter of Barcelona with him and fail to be recognized. However, until this book, until now, his name has been virtually unknown in the English-speaking world. But we, Margo Baldwin, John Barstow, Lynn Margulis, and Dorion Sagan, in our role as editors, suspect that the astounding ability of Punset to draw out the personalities of serious scientists, native speakers of so many different languages (nearly none his own native one), will astonish. His sensitivity, broad knowledge, and passion for the science itself gets each interviewee to describe the importance of the research in his or her own way. To us his talent is unique in the world. We want to share it.

How are the measurements of brain science related to our subjective feelings of happiness and despair? What aggressive or sexual behaviors documented in chimpanzees in their native habitats are just the same as ours (we talking, dressed-up, large primate out-of-Africa animal so populous on Earth today)? What evidence, based on what observations and measurements, permits us to conclude that life exists anywhere beyond Earth? What kinds of tests could, in principle, prove that life is elsewhere in the solar system or beyond? How did this scientist become passionate about genetic control of development or that one come to spend nearly all her waking hours worrying about the rates of extinction of the magnificent mammals on the East African savanna? The book can be picked up and put down at the bedside or on the commuter train.

Why? Because of its question-and-answer format it reads like a good play, one with flexible, frequent intermissions. But it is not fiction. None of it is fiction. The actors, all of whom have been active researchers and scholars for years, with honesty and enthusiasm, play only themselves. None remotely resembles the plain-looking, lab-coated bore portrayed by the usual television drama or feature film. None of the women scientists are pedantic ugly clods with one-track minds and controlling personalities. Through captivating conversation, Eduardo, always the nosy interlocutor, captures the breadth of intellectual interests, general competence, liveliness, and ability to explain why one studies ferns, another chromosomes, and a third the small probability of enormously important events.

The reader needs no science background at all. Both those who have always loved books but hated science and those who have always loved science books will be astonished to see the personal side of the unique personalities. All of us readers who eschew hype, who try so hard to avoid propaganda and marketing ploy to see Nature as she really is will put down *Mind, Life, and the Universe: Conversations with Great Scientists of Our Time* only when we finally finish reading it to the end. Breathless and enchanted enough by its truths, we may be even more delighted by the warmth and the humility of its truth-seekers.

The lively and idiosyncratic comments following the epigraphs at the beginning of most of the interviews or included within the transcript of an interview derive from Eduardo Punset's introductory remarks in the Spanish version of the original book, usually his television program *Redes*. Some of these interviews and commentaries occurred after the *Cara a Cara* (*Face to Face*) Spanish book was published.

Her goal has been to bring the gloriously honest and earnest activity of original science, scientific research, and passionate investigation to the reader from the very few here. Which few? These thirty-seven interviewees are the extremely original and dedicated men and woman who play by the rules. All of the interviewees chosen for inclusion are people known personally to Eduardo, to Lynn, or to both. To all of them the idea of scientific fraud, overstatement, or hype, say, for the purpose of accumulation of personal fame, wealth, or power is ludicrous. Why? Because their joy is in the journey of inquiry itself rather than to any final destination. Excitement for them is found in the scientific traveling life, the life of study of Nature, the observation and documentation of exploration and intellectual activity. It is a democratic excitement of discovery, one to which you are hereby enthusiastically invited.

PART ONE

PEOPLE PRIMATES

Culture before Humans

INTRODUCTION

When faced with the extraordinary diversity of life on Earth—from 3.5 to 100 million extant species according to E. O. Wilson's best estimate (see chapter 5)—I find it more than useful to divide all life-forms into two huge groups: prokarya (or prokaryotes, all the bacteria including the archaebacteria, or "archaea") and the larger organisms composed of nucleated cells, eukarya (or eukaryotes). The latter, all of which evolved from integrated symbiotic bacteria, of course include animals and plants. But eukarya also includes fungi (yeasts, molds, mushrooms, puffballs, and their kin) and my favorite group of beings, protoctists. Familiar protoctists include all the algae, the slime molds, water molds, and another fifty or so major groups. An estimated 250,000 species—all the extant amoebae, ciliates, foraminiferans, diatoms, brown seaweeds, and many other "water neithers," large and small organisms, neither animal nor plant—are included among the Protoctista. No matter what species about which we talk, the vast majority of its relatives, inferred from their fossil remains, are extinct. Probably 99.9 percent of all species ever to live on Earth, with paleontological records that extend nearly 4,000 million years into the past, are extinct. Apparently very recent, that is, from 3 million years ago until the present, 20 species of fossil humans all are included in this tally: *Homo erectus*, *Homo habilis*, *Homo neandertalensis*, *Homo ergaster*, and so forth, are among the permanently vanquished. Only we *Homo sapiens*, in prodigious numbers, remain.

Folk classifications around the world, however, usually classify organisms as either edible, poisonous, predator, pest, or scum. Like kindergarten children, the living world is seen in relationship to our families and us. A more observant popular idea is still in vogue: For years all nonhuman life was forced into two "kingdoms," either Animalia or Plantae. Indeed, in most folk classifications this dichotomous scheme persists as an unquestioned truth. With one caveat in our self-proclaimed "civilized world," three great great groups of the living are acknowledged: "animal," "plant," and "germ."

This literate world is ignorant. The word *germ* to the naturalist is akin to the word *weed* to the gardener or farmer, a value judgment that refers to rejects: the unwanted bacterium, fungus, or plant.

I have often tried to explain the two super kingdoms (or domains: Prokarya vs. Eukarya) and the five-kingdom system of classification (Monera, or Bacteria; Protoctista; Animalia; Fungi; and Plantae) and show why the "animal or plant" prejudice is obsolete. I emphasize that the threefold "animal-plant-germ" scheme has become the accepted, unstated, most popular view. Critics have emphatically corrected me in public several times. No, they claim, most of us have a four-part classification of life in mind: "animals, plants, germs, and people." Why does this view prevail and how do I know? Because, my critics claim, most don't believe we are animals at all. They always say "people and animals" when, of course, they mean "people and other animals."

All of the scientists in this section of this book know that people are not only members of the Kingdom Animalia but that we are classified as craniates (hard heads), chordates (dorsal hollow nervous systems), vertebrates (having a vertebral column, that is, we are animals with backbones), placental mammals (hair, mammary glands), primates (because of many aspects of our skeletons), catarrhines ("old world,"that is, "African-Eurasian" monkeys and apes), and anthropoid apes.

This classification, as Charles Darwin emphasized, represents our "genealogy." The embryo-forming animals evolved first, in the late Proterozoic eon from protoctist ancestors. The craniates evolved later, in the Proterozoic eon. A simplified geological-time rock scale is depicted in Table 1.

The ways in which we *Homo sapiens* would be much happier mammals become clear in Eduardo Punset's interviews. We need to face the details of our animal history, of the behaviors and the cultures that preceded the evolution of recent men, women, children, and especially babies.

Table 1. Geological Time (not to scale).

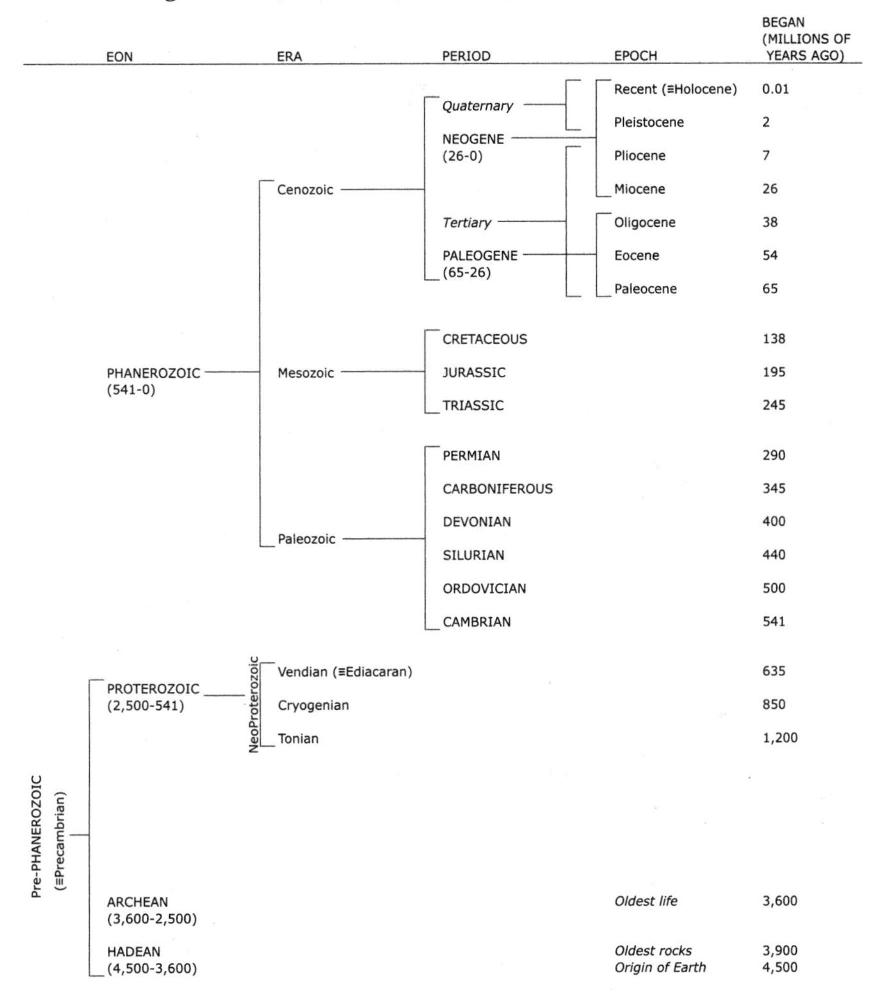

EON	ERA	PERIOD	EPOCH	BEGAN (MILLIONS OF YEARS AGO)
PHANEROZOIC (541-0)	Cenozoic	*Quaternary*; NEOGENE (26-0)	Recent (≡Holocene)	0.01
		Quaternary; NEOGENE (26-0)	Pleistocene	2
		NEOGENE (26-0); *Tertiary*	Pliocene	7
		NEOGENE (26-0); *Tertiary*	Miocene	26
		Tertiary; PALEOGENE (65-26)	Oligocene	38
		Tertiary; PALEOGENE (65-26)	Eocene	54
		Tertiary; PALEOGENE (65-26)	Paleocene	65
	Mesozoic	CRETACEOUS		138
		JURASSIC		195
		TRIASSIC		245
	Paleozoic	PERMIAN		290
		CARBONIFEROUS		345
		DEVONIAN		400
		SILURIAN		440
		ORDOVICIAN		500
		CAMBRIAN		541
Pre-PHANEROZOIC (≡Precambrian): PROTEROZOIC (2,500-541)	NeoProterozoic: Vendian (≡Ediacaran)			635
	Cryogenian			850
	Tonian			1,200
ARCHEAN (3,600-2,500)			*Oldest life*	3,600
HADEAN (4,500-3,600)			*Oldest rocks*	3,900
			Origin of Earth	4,500

— 1 —

Intelligent Life on Earth?

Interview with Nicholas Mackintosh

Greater complexity does not strictly mean greater progress.
—Nicholas Mackintosh

If Jordi Sabater Pi (see chapter 4) believes one must respect the evolutionary line of animals, rather than "trying to teach them how to sweep for us," Nicholas Mackintosh believes that it is increasingly difficult to differentiate the evolutionary process in humans from that in other animals. Mackintosh has demonstrated that in the cognitive processes of both animals and humans are characteristic structures of what is called "higher intelligence" that, on occasion, we clearly share.

Nicholas Mackintosh, a research psychologist, has recently retired from

Nicholas Mackintosh is Emeritus Professor of Experimental Psychology at the University of Cambridge.

Cambridge University in England. His investigations on associative learning reveal that the learning process in humans does not differ greatly from that of other animals.

Eduardo Punset: Animals are not like machines, you say, but rather their learning process is very sophisticated and governed by principles that sometimes also govern human learning.

Nicholas Mackintosh: The human animal.

E.P. Yes, human animals. Do you mean to say that such principles are equally applicable to pigeons as to humans and other mammals?

N.M. Yes, the principles that govern learning in pigeons are identical to those in rats, monkeys, children, or you and me. These principles are actually much more complex than what behavior experts believed fifty or a hundred years ago. These principles indicate that other animals solve very complicated problems. They understand their world and predict consequences.

E.P. And predict what will happen, although vaguely, somewhat like human beings?

N.M. To foresee the immediate future is one of the great obsessions in evolution.

E.P. You differentiate between scientists that argue that animals learn via instinct from those that believe that learning is a gradual process. How do you interpret animal behavior?

N.M. Following the principles established by John Watson and the behavior experts, I believe that animals are born with few reflexes or basic instincts. They gradually learn a particular behavior. Children are born with even fewer instincts and reflexes than other animals. We had always thought that learning was simple, but I believe that we must understand that the learning process of all animals is much more complex than what Watson and B.F. Skinner ever imagined.

E.P. What has been discovered? The importance of genetics in learning?

N.M. Yes. Genetics is very important because the learning process is conditioned by the need to solve certain problems that facilitate understanding about how the world is structured, even though not all learning systems comply with this evolutionary need. The learning system of mammals based on certain principles and laws is very limited.

E.P. It is pure genetics, isn't it? Genetics plays a major part in how people learn and memorize, right?

N.M. Yes, the process of social learning, which teaches animals to predict the future and understand their environment, a fairly universal process, is also genetically limited. When psychologists talk about the genetic roots of behavior, they refer to the innate differences between people. When they speak about "genetics of intelligence," they try to explain why some people apparently are more intelligent than others.

E.P. Old taboos prevail in "intelligence research." An experiment among school-aged children (white, black, and Asian) concluded that Asian students were more intelligent than whites, whites more intelligent than the black students. As a psychologist you have not analyzed racial differences, only common characteristics. What do you think about this study?

N.M. First, I must say that processes of learning and memorization vary much less between different ethnic groups than other measured cognitive processes (e.g., intelligence tests or intelligence quotient, or IQ, tests). There are more differences between the various intelligence coefficients than between learning processes. If you subject different ethnic groups to intelligence tests, generally speaking—at least in the United States—the majority of blacks receive lower scores than do whites, who in turn get lower scores than Asians. But the results of simple learning tests show that the three groups learn in a similar way.

E.P. Is there an "intelligence genetics?"

N.M. Yes, certainly. As a result of the different outcomes from intelligence tests, one can ask if intelligence is genetically coded. This idea creates a great deal of controversy, but I believe that the difference in results between blacks and whites is not genetic. It has been shown that the difference between whites and Asians is genetic! This difference does not refer to the entire intelligence coefficient (IQ), but rather to several specific parameters that measure spatial visualization skills—the ability to perceive shapes in space. The tests, for example, measure the perception of a rotated shape, or the identification of an object from many different angles. East Asians score much higher than do whites; even if white families had adopted the Asian children at young ages (six months), they still do better than white children. The possibility exists that this difference is genetic in origin.

E.P. Earlier you mentioned some interesting conclusions from your experiments. Apparently humans are very geometric beings.

N.M. We live in a geometric world, full of circles, squares and lines. . . .

E.P. I always tell my architect friends that we live in modern apartments like the caves in which early *Homo sapiens* lived. Any apartment in New York, London, or any other modern city has very geometric shapes, and almost the same cavelike design: three walls and a window. . . .

N.M. But not in Barcelona, because Gaudí never designed a single straight line.

E.P. Rats hate geometric shapes. They become very sad if you enclose them in a geometric place.

N.M. True, but they almost immediately learn about the space they live in, especially if you hide food in a corner, they then differentiate the shapes of their environment.

E.P. You believe that not only do they learn about the shape of their environment, but that the "idea" of the geometry in which they live superposes their own "innate" ideas.

N.M. True, and very surprising. Actually, we still don't understand why. If rats are placed in a big rectangular box and food is placed in one of the corners, rats will look for the food in that corner, but also in the diametrically opposite corner. In a sense it is an equivalent corner. If food is indicated with a signal, the rats will ignore it and continue to search for the food in the two corners. Even though there are hints that indicate which corner the food is in, rats follow the geometry.

E.P. But they are flexible, aren't they?

N.M. Yes. In the end they do learn because they are not stupid. Yet it is surprising how important they consider geometry is above other signs. It is most strange because their natural environments are not especially geometric.

E.P. Nicholas, as a result of your reflections on intelligence and learning in animals, it occurred to me that I, who teach classes at the university, know almost nothing about human learning, even of my own students. Has there been more research on the learning process in animals other than humans?

N.M. Yes. You are right. Although I am actually a psychologist, I specialized in learning in [other] animals. I believe that educational psychologists have probed less deeply into the educational process of humans than scientists have into learning in other animals.

E.P. Perhaps this is due to the fact that it is far easier to experiment with animals other than with humans. Or might it be a question of ideological influences? Theoretical bias?

N.M. A major part of psychology is overinfluenced by trends, I think. The latest trend determines how children are taught to learn, read, write, and solve mathematical problems. The trend leads education by the hand, and hence the discredit of educational psychologists.

E.P. The issue is complex. Back to animal behavior, I think it is extraordinary how pigeons use the sun as a compass. Research psychologists tell us that use of the sun as a compass is not enough to find one's way back. What do they actually mean by "not enough?" Apparently this also applies to bees.

N.M. Pigeons use the direction of the sun to synchronize their internal clock to head north, south, east, or west. Imagine that I blindfold you and I leave you 60 kilometers away with a compass. How will you find your way back to Barcelona if you don't know in which direction it lies? If you know that it is south, the compass will help you find south, but if you don't know the direction you will not know how to get back even if you have a compass. However, if you take a pigeon and release it 130 kilometers from where it lives in a place it has never been to before, after two or three minutes of flying it will start to head in the right direction in order to return. They believe this ability is based on experience and training, which is why very young birds don't know how to return, because even though they have a compass, they have not yet drawn up their map. When they are trained on short flights, pigeons identify the area over which they fly. For example, they hear the sound of waves breaking on the shore, or smell the city from 130 kilometers away. Pigeons find their way back if they previously have undergone training flights. They will have learned that if a particular sound at the seashore is louder and it is in the north, if the bird is farther north and the sound is even louder, it means that the bird only has to head south to return home.

E.P. I am not surprised you wrote that it is difficult to demonstrate that the most advanced forms of intelligence are only found in higher animals.* What do you mean exactly?

N.M. What do I mean? Simply that certain forms of intelligent behavior are found in birds as well as in monkeys.

E.P. Or in human beings.

N.M. Yes, some forms of intelligence are found much more abundantly in humans, but I was referring to nonhuman animals. Some bird species like ravens, magpies, and parrots, solve very complex problems beyond the ability of many simple primates and other mammals. Some ways of reasoning, the capacity to establish relationships between events and the comprehension of analogies, are ways of learning characteristic of some primates, especially monkeys, but also of some birds, and not rats, for

*Darwin admonished us not to use "higher" or "lower" animals or plants with the inevitable value judgement. He noted that all organisms alive today have survived and therefore are equally "high." (Letter to Hooker, in *Darwin: The Survival of Charles.* Ronald W. Clark Random House, NY 1984)—LM

example. The pigeon, like many birds, has a great ability to orient over long distances, but this is such a specialized ability that we tend to think it does not form part of intelligence. Many animals have the ability to navigate around the world. An ant can leave the nest and go forward in a zigzag pattern, find some food, and return in a straight line. Although it came out in a long zigzag course, it knows the direct way back.

E.P. I imagine that you agree with Stephen Jay Gould's idea that humans "are not headed toward something bigger and better—"

N.M. Yes, there is no aim of progress in evolution.

E.P. What do you think of this idea after so many years of investigating evolution?

N.M. I agree with Gould to a certain point. It is incorrect to maintain that evolution is a continuous progressive process that culminates in humans. To think this is both anachronistic and anthropocentric. I have always tried to refute it by arguing that higher intelligence is not exclusive to only those animals closely related to human beings. Rather it is also found in animals that are only distantly related, such as pigeons and ravens. Evolution in this sense does not mean progress. It is not necessarily related to progress. Evolution has evolved greater complexity, without a doubt. Generally speaking, evolution does not lead to lesser complexity over time. But there are examples of animal degradation where they lived in caves or underground and lost their sight. Because competition makes complexity necessary to resolve problems, evolution produces changes and weaves a greater complexity. But greater complexity does not strictly mean greater progress.

— 2 —

Stressed Chimps

Interview with Robert Sapolsky

For the typical mammal, stress is induced by another who is intent on eating you in the next two minutes.
—ROBERT SAPOLSKY

This transatlantic and transcontinental telephone call took two hours and filled twenty-five pages in two or three rough versions. My co-editor (LM) reluctantly cut it, smoothed it out, and made it look like the others in size and style. But Sapolsky is an original. I recommend you read his work; he has much more to say than we could fit in here!*

Robert Sapolsky is the John A. and Cynthia Fry Gunn Professor of Biological Sciences and Professor of Neurology and Neurological Sciences at Stanford Medical School.

---

*We have solicited representative accessible readings, limited to two, from each interviewee. The Readings begin on page 341.

Eduardo Punset: You are somewhere near Stanford in California, and I am here in Barcelona—a nine-hour difference.

Robert Sapolsky: It is about seven thirty in the evening.

E.P. When it is seven thirty here what time is it where you are? About ten o'clock in the morning?

R.S. Yes, ten thirty.

E.P. It is wonderful to speak to you. I tried to visit you, but a series of minor events delayed my trip to San Francisco. We'll try to overcome our technological constraints.

Let's talk about stress. When a stressor throws my body out of what you call "allostasis" (which for now I'll refer to as balance), my brain is my "master gland." Without my awareness it automatically secretes hormones to restore the balance. My brain tells my finger to move. I understand that. But how can my brain, all by itself, start this complex process of restoration of my body from the effects of stressors on my health?

R.S. Sometimes the process is very complex. You sit and think: "Okay, how many days now until my deadline—Oh, my God! It's only four days!" BOOM! Suddenly off you go with a stress response. Sometimes your response is due to what is called the "preconscious." Preconscious processing in the brain refers to a fear response, elicited by something of which you may not even be conscious. Those with post-traumatic stress disorders are not even consciously aware of fear-evoking stimuli that enter their environment. "Aha, this is a similar voice as the person who did that to me," or "Aha, this is the dark alley scene I was in when that happened to me." A preconscious stimulus and, consciously aware or not, suddenly the heart races, one is flooded with the sense of panic.

E.P. Fantastic. Just by thinking, if I find a lion around the corner, or just by imagining that I might find a lion around the corner, the sophisticated process kicks in, described in your marvelous book, *Why Zebras Don't Get Ulcers*. Not only do hormones prepare me to face the lion that I have not seen, but they also inhibit physiological processes geared toward the long-term, such as immunity or sex drive. I feel less sexually driven because I imagine a lion around the corner. Isn't that incredible.

R.S. If you think about the way any animal becomes stressed, rather than we Westernized humans, it makes sense. For the typical mammal, stress is induced by another who is extremely intent on eating you in the next two minutes. Or perhaps you are intent on eating someone else in those two minutes. How your body responds under those circumstances is brilliant. It is exactly what is needed. Energy is mobilized from storage sites to handle whichever muscles are exercising. Blood pressure is increased to deliver that energy faster. As you said, long-term building projects turn off. A lion chases you, then you ovulate at some later time, or hit puberty on another day. Don't grow, or even digest your meal, you'll make your antibodies tonight, if you survive until then. Shut down nonessential systems. Of course, the trouble is that, as very psychologically sophisticated primates, we activate the same stress response when we think about an emotion, psychological state, a memory, or a past experience. We image what might happen 30 years from now, or may never happen. The exact same stress response is turned on. The central concept is that if you activate the stress reaction for three minutes and run to save your life, it is wonderful. If you chronically turn on the stress for psychological reasons, more likely you will get sick.

E.P. Is it true that 60,000 years ago the stress reaction was more or less the same as it is today? The consequences are the same. Is the response as useful today as it was 60,000 years ago?

R.S. Great! First off, I suspect the stress response is closer to 100 million years old! I predict (scientists love these sorts of untestable predictions) even dinosaurs had very similar hormonal stress responses. Lizards, fish, birds, all have the same very ancient response to a physical stressor. As you say, 60,000 years, or maybe even longer, ago, we became sophisticated enough as primates to turn it on for psychological reasons. In evolutionary time, 60,000 years is short. We now have a wonderful system that makes sense to a real mammal if we are running for our lives or even for the bus. But it is a disaster if instead we sit there thinking, "Oh, my God! I'm going to die someday! What's happening with global warming? How can I pay bills by the first of the month?" Stress from psychological anticipation is not good.

E.P. Is there a danger of overreaction to a stressor? Might that explain why so many wrong deeds are committed?

R.S. As we look at the range of the most common psychiatric disorders they are diseases of people who do not deal well with stress. Major depression is when in the face of stress and challenges one immediately says, "I give up, I won't bother to cope. I am helpless, it is hopeless. . . . " Completely subjugated, they don't try. More relevant to your question are people with anxiety disorders. When stress occurs they do try to cope by doing seventeen different things at once. They don't stop trying no matter the outcome. They behave the same way over again in a constant psychological state of emergency, even in the absence of stressors. Also critical to making sense of some of the most miserable aspects in human social experience is that one of the best stress-reducing actions that animals, including us, do is to make someone else miserable. Displace our aggression onto them. Many of us avoid ulcers by stressing and giving them to someone else.

E.P. One way to reduce our anxiety or maligned sadness is to bark and make somebody else unhappy. Robert, let's discuss what I single out as one of your greatest contributions to knowledge. I know many scientists also look, but you started this. I just invited members of my audience to count their heartbeats after first trying to relax. Then I just reminded them that, with each beat, some part of their life goes away, irretrievable as they come nearer to death.

R.S. Very nice.

E.P. Suddenly the heart beats faster. A clear example to me of one of your greatest contributions: the power of imagination. I become almost ill, because something happens in some part of my brain.

R.S. Very few of us will die of smallpox, malaria, or yellow fever. What we die of are "Westernized" diseases. The body slowly falls apart over seventy or eighty years: diabetes, hypertension, cancer. I believe the single most important sentence in all medicine these days is, you directly affect how your body works, yes, with thought, with memory, and with emotion. This is unrelated to survival of anthrax or smallpox, but has everything to do with how the body handles diabetes and heart disease. In smart social primates like us, thought, emotion, and the ability to dramatically change how our bodies work and respond is central to who among us remains healthy and who gets sick! If you sit there and think: "Oh, my God, I'm going to die!" or "Oh, my God, I forgot to pay my electric bill!" what happens is exactly what you need as a normal mammal: a sense of urgency. If a carnivore is trying to eat you or you are hungry, hunting a deer, or running for your life across the savanna, what happens to your body is wonderful! You secrete stress hormones. They deliver energy into your bloodstream. Your muscles work better, your heart rate increases, so does your blood pressure. All long-term physiology cuts off: no time to have puberty, or to thicken your uterine walls.

E.P. No sex.

R.S. Absolutely. You will make sperm another time, not now. No. Run for your life. This makes perfect sense to a normal stressed mammal in a very short period of absolute physical terror. We dwell on alarming facts: rain forests disappear, the ozone layer is gone, the planet is warming, we will die some day. We activate the exact same stress response. The central tenet of the whole field of study is that if stress formerly essential for purely psychological reasons is chronic, it will make you sick.

E.P. Did you not measure the hippocampus, that particular region of the brain, and find that chronic stress devastates it and some may lose 20 percent of it? Can this be due to just sheer thinking?

R.S. Yes, very disturbingly. Only in the last twenty years or so has the fact been established that stress, in addition to effects on memory loss and rise in blood pressure, can badly affect the brain! Much more substantially, stress actually kills neurons in the area of the brain called the hippocampus, one that has much to do with learning and memory.

E.P. Yes.

R.S. The area damaged in Alzheimer's disease is the hippocampus. We have seen for twenty years that it can be damaged thanks to stress hormones. Does this apply to humans? The literature suggests that people with major clinical depression that has gone on for years have elevated levels of stress hormones, hydrocortisone, consequent shrinkage of the hippocampus. Memory problems ensure the longer the depression, the more shrinkage. Some of our brains age faster than others. I have spent the last twenty years in study of the hippocampus. Perhaps I chose the wrong area of the brain! The hippocampus is important, but much more interesting is the most recently evolved part of the human brain, the area called the frontal cortex. We have more of it than nonhuman mammals. The frontal cortex disciplines us: it is involved in gratification postponement.

E.P. Planning?

R. S. Yes, planning. The frontal cortex keeps us from murderous behavior. You are at dinner with a guest and they do not finish the meal. You want to say, "You pig, you ingrate, how could you not finish?" But you don't. Why? Because the frontal cortex modulates the much more impulsive aspects of our behavior. No surprise that kids do not have a whole lot of frontal cortical development. The full development does not occur in people until we are about twenty to twenty-five years old.

E.P. I see!

R.S. Let us look at what happens in the brain: when you sleep your basic metabolic rates, energy rates, lower throughout the brain. You enter deep, slow-wave sleep. Around ninety minutes into sleep, you enter the first dream stage. What happens? Parts of your brain, areas related to sensory processing, not only become ever more active, they become more active than when you are awake. You dream, visual and auditory stuff. Except for the frontal cortex, the rest of the area doesn't change at all. In the frontal cortex metabolism decreases. It goes down to zero when you dream. Suddenly you dream that you are floating in air. . . .

E.P. No inhibitions?

R.S. You're a dolphin, you're—

E.P. No discipline?

R.S. Zero! And this, in a sense, is why dreams are dreamlike. Maybe Salvador Dalí had no frontal cortex, or he turned his off when painting. Surrealism involves disinhibition of the world's linear features. Watches do not melt. When you dream you enjoy that same freedom from logic, freedom from discipline. This is why dreams are dreamlike, because your frontal cortex stops working.

E.P. And what about your chimps, or the monkeys you work with, the baboons?

R.S. They have very few, maybe three frontal cortical neurons. They are some of the most impulsive animals on Earth. Here's a good example: Over and over and over, baboons are smart but they are so impulsive. They can't help acting stupid and aggressive at the worst possible moments. A baboon has a better chance of killing a gazelle if it hunts by itself than if it hunts in a group.

E.P. Why?

R.S. Why? This whole group, it's four or five males. They search, and they chase the gazelle. Everything is fine. They come closer. I don't know what the baboon is actually thinking, but as far as I can tell, at some point he says: "Well, here I am, running as fast as possible, and here's this guy running as fast as possible. . . . Why? I don't know. I can't remember. But if he's running after me so fast we must be in a fight." He suddenly stops, turns around, and the two of them go rolling.

E.P. And the gazelle?

R.S. And the gazelle runs off! They're very impulsive. They have very little frontal cortex. Humans spend forty years to save money for retirement. Baboons could never do that.

E.P. Thanks to our neocortex, larger than in other species, we plan to a further extent, and we have more discipline. But are we more intelligent? This reminds me of another fascinating story in your book about a microscopic worm, a parasite that enters a host, a mouse, with the hope of being taken into the only place where he can reproduce, into a cat. So, when in the mouse, the parasite manages (please correct me) to interfere with the neuronal processes of the mouse. The mouse loses its visual panic of cats. The mouse becomes a tender cat-loving rodent, so that in the end the cat eats the parasite-ridden mouse, where the parasites reproduce happily. This parasite, isn't it intelligent?

R.S. Terribly evolved! We humans have hundreds of billions of neurons. We can invent microwaves. We say we are totally amazing, but these parasites, bacteria, viruses, they enter the nervous system and change the behaviors of their hosts. Look at the rabies virus. It makes the mammal rabid! It enters the brain and makes it rabid so that it tries to bite someone. Rabies viruses shed thousands of viral particles into the saliva to pass themselves to the next victim. The rabies virus knows more about the brain, the causes of aggression, than hundreds of neuroscientists do!

Women, when they are pregnant, suddenly worry about cat litter, because toxoplasma (the protist *Toxoplasma gondii*) is in the feces that the rat or mouse ate. Toxoplasma's evolutionary challenge is: I'm inside this rat or mouse but how do I get inside the cat's stomach? Some parasites destroy the muscles so the host animal can't escape from the predator! But toxoplasma enters the brain. You know every mouse and rat on Earth dislikes the smell of cat. A laboratory white rat, four hundred grandparents' generations away from any cat, given the smell of cat instinctively moves away from the cat odor. With toxoplasma inside its brain this animal suddenly loves the smell of cats and goes right up to them! Everything else about the rat works just fine. Their ability to learn tasks, their social behavior, is normal. But somehow toxoplasma knows how to undo the wiring of a basic innate rat phobia. How does this toxoplasma work? No one has a clue! But just as you sit in this TV studio and say, "Aren't we humans marvelous?" you must face the fact that toxoplasma rewires rat brains with a finesse we people lack.

E.P. Some of our characteristics, special to us, seem to confuse us, rather than to make life easier. One of your examples, the limbic system, the brain part associated with emotion and feelings, seems to work more quickly than the autonomic nervous system. Please give me your example related to social relationships.

R.S. I love this example. It is certainly relevant to my own life and those of many other people. So I'm with my "significant other," my partner, and I've done something just terrible. I've just been a jerk. We have a fight. At some point I realize, actually I've been ridiculous and I apologize. But she begins to be furious at me. Then she accepts my apology. I begin to think, *Good, that's the end.* And then what? Suddenly she remembers some jerky thing I did twenty-seven years ago. Something I forgot about years ago. She suddenly remembers it and wants to argue it all over again. "I can't believe you did that back in 1980." What is this? I think an explanation is related to how we feel emotion. How we decide what the emotion is. What is logical? The limbic system, the part of the brain with emotion, decides something is upsetting, exciting, arousing, whatever. It tells my body, and my autonomic nervous system kicks into gear. My heart beats faster. I sweat. That's what it is.

A hundred years ago, or so, William James discovered much about how physiology and psychology fit together. He suggested that part of what our brains do is they say, "What is happening in my body? If my heart is racing, I must be feeling a strong emotion. If my breathing is regular and slow, I must be feeling quite calmed." That might sound crazy, but he was right. Part of how you decide what your emotions are, and how strong, is in the brain. Your brain monitors your body and picks up autonomic physiological signals. It decides if you are angry, or if you're really angry, and it generates a loop. The key point, as you mention, is that something gets you upset, makes you angry, and activates your limbic system. A few seconds later then your autonomic nervous system activates as well. Finally you apologize. The fight is over. You realize you were a jerk. It's true. Great. It is settled, but the autonomic nervous system returns to normal more slowly than does the limbic system. The fight is over. Finished. She's not allowed to be angry anymore because I apologized. Yet my heart is racing, and my stomach is clenching. William James comes back now to ruin my life. My brain says, "Well, let's see, but my heart is still racing and my stomach is still clenched, I must be upset about something if that is what my body still does. Is she upset? No, no, I apologized. It must be over with!"

E.P. It all happened years ago—

R.S. Yes—it was over twenty years ago. "But we need to talk about it again, right now!" The autonomic nervous system turns off more slowly. What is interesting is that, on the average, it turns off even more slowly in women than in men. This scenario apparently explains a familiar experience. Just when I thought it was over with, it is not over—twenty-seven years later.

E.P. Robert, I have a female dog, called Pastora. When I was writing a book last summer on what science says about happiness I realized that my dog was very funny in one way. Her dish is outside, on a small terrace. When I went to pick up the dish on the terrace she started jumping around. I could hardly walk. Until I get to the dish, then I go to the kitchen with the dish, and Pastora will run around like that and I say, "Look, Pastora, calm down, otherwise I can't even get to the kitchen." So I go to the kitchen, fill this dish with dry dog food and occasionally a bit of ham. She will wait there, you know? If I take more time she will bark. The same theatrical performance begins again on the return to the terrace where she eats. For years I wondered what was the matter. Because when I give her the dish, sometimes she won't even eat. The whole animal changes. I was writing this book, *The Happiness Trip* (Punset, 2007), and I said to myself, "Look. Listen. The great happiness seems to be in the waiting room. It's a matter of expectation." Once she gets her food, it's over. When I went to your books, I found the same sort of reasoning that you also write about: the anticipation of pleasure is where the pleasure is. Explain the relation to dopamine or other neurotransmitters.

R.S. Yes, dopamine.

E.P. It's dopamine I think your researchers have found. We secrete dopamine. Does my dog, Pastora, also secrete dopamine before eating, not while she eats, but in anticipation of her food?

R.S. Yes, your explanation is perfect. We did talk about dopamine one time. Dopamine is about pleasure. It is the reward. The explanation is about the reward; it's the anticipation of the reward. We train a laboratory rat that when a light comes on in his cage, it means to go over and press the lever five times until he receives some food. The first time the rat gets the food, up goes his dopamine level. But after a while, when does the dopamine level rise? Not when the rat receives the food, but when the light comes on! In anticipation.

E.P. Like Pastora.

R.S. Yes, exactly like Pastora. The rat sits there and says, "This is great! This is great! I know that light, I know where the lever is, and I can hit it. Look, this will be fabulous. I'm on top of it, I'm in control!" The pleasure is all about the anticipation. The dopamine rises with the anticipation. OK, so you are correct. I recall that a friend of mine in college had a most cynical view of nearly everything. He said, "A relationship is the price you have to pay for looking forward to it." This fits. He already knew what dopamine does.

Here is an even more interesting discovery from an extensive study a couple of years ago. The rat presses the lever and will get the reward. It is all worked out, except that now you, the investigator, respond differently. Instead, the rat gets the reward only 50 percent of the time.

You introduce uncertainty. And what you see, right after the lever is pressed, is a rise in dopamine, a huge rise, larger than ever before seen in brain chemistry studies. It stays up until the rat finds out if he will get the food or not. In other words, when you get just the right amount of "maybe," or uncertainty, it is even better than "here it comes definitely!"

E.P. God!

R.S. They show in these studies that the uncertainty of 50 percent creates the largest rise. With 25 percent, 75 percent, there is not as much of a rise in dopamine. "No, maybe it's not going to come! I'm not sure; yes, today I feel lucky!" It is anticipation. People who study the psychology of stress always emphasize that when we feel like we have no control, the feeling is very stressful. But here laws of maximal unpredictability feel great. What's the difference? Because we're optimistic in this benevolent setting. We say to ourselves, "This will probably work out OK. I will be fed or I will have sex. If not this time the next time . . ." and then maybe it is just the greatest thing you can imagine. And when this study was published many people commented on how much it teaches us about addiction and gambling.

— 3 —

Nearly Human

Interview with Jane Goodall

Grooming . . . also has a social function.
—Jane Goodall

What fun it has been to interview Jane Goodall and talk about her unique experiences among the African chimps.

Eduardo Punset: Prosopagnosia (face blindness) is a neuronal irregularity where those who suffer from it forget faces. A surprising aspect about your marvelous book is that although you forget human faces, in the case of the chimpanzees, with whom you met and lived in Gombe, you remember their faces well. How do you explain this?

Jane Goodall: Fortunately I do not suffer a severe form of this strange condi-

---

Jane Goodall is famous for her 45-year study of chimpanzees in the wild, primarily in Gombe Stream National Park in Tanzania.

tion. The first time I see a human face, it is possible that I will not recognize it, but after a while I do not forget it. It is the same with the chimpanzees. It was more difficult for me than for other people to identify them by their faces, but once I recognize them I do not forget them again. It just takes me longer to recognize them in the first place.

E.P. You are almost a scientist, Jane, in the sense that scientists tend to question nature rather than people. Ever since you were a young girl you began to question nature. You've written this wonderful anecdote when you tried to discover where the hen's egg came from. Please tell us about it yourself.

J.G. My first memory is when I must have been about four and a half years old. We lived in the center of London. We went on holiday to the country. I was always a great lover of animals, and was surrounded by cows, pigs, and horses. I had to collect the hens' eggs. The cruel farms of today didn't exist at that time. Hens ran freely in the farmyard, and were urged to go into the henhouse to lay their eggs. I had to collect the eggs, which were about so big. But where was the hole in the hen where the egg came out? I couldn't see it and I asked everybody. Their answers did not satisfy my curiosity. I climbed up to the henhouse to see how they did it, and when the hen saw me she flew away. So I climbed into an empty henhouse and I waited for four hours until a hen came in. I watched how it laid an egg. By then my mother was worried and had called the police, but it didn't matter. I had seen something marvelous. And, as you said, curiosity is the beginning of a scientist. Asking questions, not finding answers, then looking for answers for oneself. If it doesn't work the first time, try again. One has to have a lot of patience.

E.P. Patience, that's the word. Later, with the chimpanzees, you would prove that you have a lot of patience—in Gombe?

J.G. Yes, in Gombe.

E.P. The other day I talked to Nobel Prize winner Sydney Brenner [see chapter 19]. He told me he currently studies the nervous system of octopi. After he spent so long studying the genome, now he studies the brain. Sydney Brenner said something that reminded me of you. He said that he had faith in the ignorance of others, that usually they could not concentrate on a subject sufficiently to find the solution. A certain degree of ignorance seems involved in the first study of a subject. And when you were a young girl and confronted with fox hunting, it is surprising because at the end of the story you said that the only way to correct this is for other people who are not fox hunters to study the problem and find solutions. Do you believe this is true?

J.G. What I believe is that, in the same way as other multiple examples of cruelty to animals, the perpetrators of the cruelty do not really understand it. My work has always been to help people understand the true nature of animals. Foxes are very intelligent, they have extremely developed brains, and they feel emotions such as fear, and the solution is not to hunt them. It is easier to live without hurting animals. It's very easy.

E.P. Now let's talk about your adult life, when you began to research chimpanzees. You met Louis Leakey, who had found *Australopithecus robustus* fossils. He was a genius who helped you to establish yourself and enabled you to research chimpanzees in Gombe. He believed that by study of chimpanzees we could understand our ancestral origins better—

J.G. Or how they behaved. He found he could determine the physical appearance with fossils and what they ate. But he believed that through the study of their closest relatives alive today he could get a more precise idea of the social behavior of extinct hominids—if you agree that the chimpanzee is our common ancestor who dates back to around seven million years. If we find a behavior similar in chimpanzees and humans today, it was probably the same in the common ancestor in the Stone Age. This was his theory. Anthropology books often mention chimps as behavioral models of "primitive man."

E.P. And you proved that you, like when you were a child, had patience and willingness to observe. Often primatologists worry about their influence on chimpanzees, and some even teach them sounds, but the incredible thing about you is that when you were young you spent hours upon hours observing them with the patience of your childhood. When you decided that you wanted to see the world or the universe through the eyes of a chimpanzee, you couldn't. Or could you?

J.G. Sometimes it is easy to see through the eyes of a chimpanzee who looks at a mountain of delicious fruit. I get a fairly correct idea of how a chimpanzee feels when she looks at the fruit mountain. But when they sit in the top of a tree in the afternoon looking around them, although we do know that they think, we do not know what they are thinking. My oldest chimpanzee friend, Fifi, was a baby in 1960. For forty-five years she always knew when I was coming to see her. She came to greet me. She always sat more or less how we are now—or maybe a little farther away. Often all we did was look into each other's eyes. I knew that she and I were the only ones who shared memories of the early 1960s, because all the others were dead. She demonstrated the huge memory capacity of chimpanzees. But now she is gone, too. My best friend is no longer here; she died in September.

E.P. And while you are talking about them, why don't we remind our audience of their body language and other similarities between them and humans?

J.G. Nonverbal communication?

E.P. Yes.

J.G. If I look at you like this while you are eating fruit, you will probably put fruit in my hand. Chimpanzees do the same. If we are friends and greet each other we will probably hug or kiss each other. Chimpanzees do the same. If I were dominant over you, you would probably be submissive. You might want to demonstrate your attitude by raising your hand a little in a threatening way. The chimpanzees do the same. The nonverbal body language is the same for chimpanzees as it is for us. They use the same gestures and postures in the same context.

E.P. Incredible! Some say that chimpanzees spend 30 percent of their time grooming themselves and from this comes the origin of language. What do you think?

J.G. I doubt it, because while grooming you do not need language. Grooming functions to remove fleas, seeds, parasites, and so forth, but it also has a social function. It reestablishes relationships, lays foundations for new relationships, repairs difficult relationships. It is a calming activity. Young chimpanzees are also groomed. When in the course of evolution hair is lost and there is nothing to groom, the need for another friendly form of contact arises. But the origins of language are fascinating. You have the right idea: it is our oral language that differentiates us most from our closest relatives. I describe scenes in Gombe to you, and although you have never been there, I can vividly paint them with words. We can teach our children about attitudes, people, things, et cetera, not present. We make plans five years in advance.

E.P. Do chimpanzees yawn? Have you seen chimpanzees yawn?

J.G. Yes, they yawn.

E.P. Some paleontologists and physiologists from the University of Maryland in the United States have studied the function of yawning in humans and have not found anything. They say to yawn is absolutely useless, and that it is just a prehuman habit inherited from chimpanzee ancestors. What function has it in chimpanzees?

J.G. None.

E.P. None?

J.G. I always thought that yawning was related to breathing.

E.P. The volume of oxygen in the lungs measured before and after yawning does not seem to change. Why is it contagious? Is yawning contagious in chimpanzees?

J.G. No, it doesn't seem to be. I don't know.

E.P. Do talk about your important contributions to primatology. Today some no longer seem new, such as the discovery that chimpanzees also use tools. How did you find out?

J.G. It was incredible. Particularly so because, at first, there was only enough money for six months of study. We knew that at the end of this period if we did not discover anything, the project would be finished. One day as I was strolling through the jungle I saw a shadow bent over a termite's nest. I looked through my binoculars. There he was, the least fearsome chimpanzee of them all, the one whom I called David Greybeard. He was using a blade of grass as a tool to eat something. What he was actually doing was using the grass to extract the termites out of their nest to eat them. What was incredible? Sometimes he would take a branch and remove its leaves and use it to extract termites—the beginning of the creation of tools. At the time it was incredible, because until then only humans were believed capable of manipulation of tools. We were known as "humans, the manufacturers of tools." So when I sent a telegram to Louis Leakey he said, "Now we have to redefine man, and redefine tools, or accept the chimpanzees as humans."*

E.P. For many years, the search for the differentiating feature between humans and other primates centered around the use of tools—

J.G. Which was not true.

E.P. Which was not true! Then they said language and communication definitively distinguishes us, which in some ways is not true either.

J.G. Yes and no. With respect to communication, other animals, including chimpanzees, communicate marvelously through a repertoire of voices, gestures, and postures. But when we come to our sophisticated spoken language we acknowledge the fact that we can speak about many things, many of which we know nothing about, but we can communicate. We can tell our children about events that happened in the past. We can make plans for the future. You and I can discuss an idea. Each of us can make contributions. Our idea may grow and change.

E.P. And do not chimpanzees work in this way?

*If we humans were not so psychologically unprepared to accept the truths of science, no doubt the chimpanzees would be classified with us in our genus, as, say *Homo gombeensis*. They resemble us far more than many animals placed in the same genus, e.g., species of *Equus* (horses), or of *Drosophila* (fruit flies).—LM

J.G. We, from what we know, are the only creatures capable of this sophisticated communication. Chimpanzees can learn sign language or even computing language; therefore their brains are capable, but it seems as though they differ from us in this way.

E.P. Do they think?

J.G. Yes. Of course they think!

E.P. Another discovery of yours is that, contrary to our belief, they are not vegetarian. In addition to slugs and termites they eat mammalian meat. How did you find out?

J.G. We know now that they are omnivores like us. Once again, the first time that I saw this was with David Greybeard. I observed him from a mountain peak with my binoculars. He climbed a tree with something that obviously was a piece of meat. A female asked him to share it with her. Under the tree there were two pigs trying to attack him. It was obvious that David was eating a piglet.

E.P. So he was eating meat?

J.G. Yes, he was eating meat and the female asked him to share it with her.

E.P. Another discovery which surprised you? Your book is full of feelings. In this world in which nature is almost idealized, you discovered that not only are chimpanzees violent among themselves, indeed they are capable of killing and eating each other! But you also observed a "war," perhaps better called, a "skirmish" or "tribal fight," between two tribes that lasted about four years!

J.G. Yes. It was shocking. It happened after the main community under study was divided. A small group settled in the southern part of Gombe and took control of this area. The males of the most heavily populated part of the northern area progressively entered the new south part, the territory of the males in the south. When they discovered one of the males of the south in their new territory they attacked him and left him to die from his wounds. So after four years the seven males and three females disappeared. The victorious males and their females occupied the new territory!

E.P. What surprised me was your memory of this—that at a certain moment in the fight, one chimpanzee touched another one. The second made a gesture as if something were contaminated in the same way as humans do.

J.G. Yes, that was awful! The female had been repeatedly attacked, but she was not badly enough hurt to die. She stretched out her arm as if to feel reassured. What she did was touch him and that is when the male cleaned himself there with leaves. They clearly established differences between their own group and the others' group. Thus began affliction in humanity, which separates one group from another and leads to the violence of wars—the capacity we have to treat humans who belong to another group with almost nonhuman labels. Reminiscent of the way the Nazis treated the Jews. It still happens today.

E.P. This chimp violence comes from the concept of territoriality, no? This territory is mine and "anyone who enters my territory will be expelled." The psychologist and neurologist Robert Sapolsky, from Stanford University, talked about violence between humans [see chapter 2]. He said that it was probably political power, the execution of power over the poor, something so vile, so wretched that goes back in our history and persists to current times. He studies stress. In some way political power is the root of violence. He notes the poor have more heart attacks than the rich, and they suffer more rheumatism. The details are in his Teaching Company lecture series. How do you feel about this?

J.G. I cannot follow this line of thought. The desire for power, yes, characterizes important men, and important male chimpanzees. They both like power. But when the poor tend to suffer from certain illnesses, I associate this with the lack of a healthy diet, to unhealthy eating.

E.P. Yes, but perhaps this bad diet, lack of healthy food, is imposed by the rich. The idea, which you share, which I have read in your writing, is that violence is different. In both cases it is horrible. As you said, it is not the same—it is different. Sapolsky's idea is that violence in chimpanzees arises from dominance, but in humans it is different, more wretched.

J.G. It depends on the violence. The violence of chimpanzees is very similar to the war between violent gangs; in both cases they fight for territory, possessions, and females. The context in which we see aggression in a chimpanzee is also present in humans. It creates aggression in us, too. But it is different for us when money comes into play and when a political web is woven. Chimpanzees lack language sophisticated enough. Although a chimpanzee and a human can inflict the same type of aggression on their victim, I believe aggressiveness in humans is worse because in principle we understand what we do, as a whole, can inflict both physical and mental violence. Humans are capable of evil, of real and deliberate torture, whereas chimpanzees react to the happenings at that moment. They do not plan for the next few weeks. They don't plan to twist someone's thumbs, or to rape or beat someone. They do not think ahead in this way.

E.P. You mentioned violent gangs and associate them with what you call "pseudospeciation." Do these chimpanzees who separate and divide the territory become a type of new species?

J.G. When the enemy is categorized as another species, we do things to that enemy that we would never do within our same community. When chimpanzees attack members of other groups, they show the same behavior as they do when they attack large predators. They do not show such violence in fights with members of their same group.

E.P. You mean that they are not so cruel?

J.G. No, they're not so cruel.

E.P. But can't they be as cruel with those chimpanzees that had belonged to their same species for a long time?

J.G. Yes, like in a civil war. They have renounced the rights that they had when they belonged to the group that they abandoned.

E.P. That explains the type of cruelty characterized by civil wars.

J.G. Yes, civil wars are horrible. In Spain we know this just as we do in Great Britain.

E.P. Let us talk about something more pleasant! About grooming, or love between chimpanzees and their maternal instinct. Some primatologists have researched the maternal instinct of chimpanzees in the zoo. They found that in a chimpanzee it is not instinctive, that in captivity they must learn how to look after their young. Is that true?

J.G. Yes, it's true. Maternal behavior is not that instinctive. Chimpanzees learn from their mothers or from other females. Typically when a chimpanzee has a good mother she not only observes how her mother behaves with the baby, but she is allowed to look after the next baby of the mother. She is like a mother to this other baby and learns to be as good a mother as her mother was. And if she has a mother who is less careful, more severe, less caring, and less playful, the young chimpanzee will grow up to be very similar to her more thoughtless, more selfish, less caring mother or stepmother.

E.P. When a mother chimpanzee is kind, caring, and protects her baby, her young tend to be more secure and happier when they are adults. Just as with humans. Have you observed this?

J.G. Yes, like us, and in fact studies began the other way round. Based on observations made on chimpanzees, child psychologists and psychiatrists began to study the notion that early experiences are so important. This has become a real problem. When we think about everything our children are exposed to in the Western world that we call "civilized"—appalling child-care centers, mothers who work all the time outside the home, broken families—it is no surprise that there are so many teenagers with serious problems.

E.P. This was observed first in chimpanzees?

J.G. Yes.

E.P. Let us remember the great master Louis Leakey, whom you admire so much. When you told him that chimpanzees used tools, he said, as you quoted earlier, "Now we must redefine tools and men, or accept chimpanzees as humans." What is the ultimate lesson to be learned?

J.G. The most important lesson I have extracted is that at the end of the day, no divisory line separates us from the rest of the animal kingdom. We are not the only beings on the planet with personalities, thoughts, and—most importantly—feelings. Once we are capable of accepting this, we consider the ways we use and abuse this planet. So many other ordinary animals think and feel. Many ethical questions arise that do not let me sleep at night. It is particularly tragic to observe these ambassadors, the chimpanzees, because they are the ambassadors of the animal kingdom who tell us that they too have feelings and form part of our world. Yes, they are a part of us. They are becoming extinct, right now as we speak. This is very sad. Only one hundred years ago in Africa there were a million of them, and now there are only 150,000. This is why I had to leave the research with my students and dedicate my own time to campaign to raise the alert about the problems in Africa. The jungles are disappearing; the wild animals are killed for commercial reasons—the bush-meat trade. This is why I feel passionate about a new generation of young people emerging that is more vigilant than we have been. Our institute's program, which extends to more than ninety countries across the world, is based on these scientific truths.

— 4 —

A Question of Degree

Interview with Jordi Sabater Pi

The time will come when our attitude will
change, but by then there will not be any of these
marvelous mammals at all left in the wild.

—Jordi Sabater Pi

Jordi Sabater Pi has always avoided falling into the same trap as those who believe that animals should imitate our evolutionary line in order to progress in their own.

"They are different and so it is pointless teaching them how to sweep the floor," he mutters as he contemplates the chimpanzees and gorillas that he has studied ever since his youth in Africa and at the Barcelona Zoo.

Jordi Sabater Pi is Emeritus Professor of Psychobiology and Ethology at the University of Barcelona.

Sabater is one of the founding fathers of ethology and primatology in Spain. He is one of the world's first pioneers in chimpanzee and gorilla field studies. He lived in Africa for thirty years, and his works, published in a number of popular science books and journals, and his beautiful drawings, give us a profile of this great naturalist.

Eduardo Punset: Gorillas have a bad reputation, don't they?

Jordi Sabater Pi: A very bad reputation. It traces back a long way to the last century, to a French-born American explorer called Paul Du Chaillu, who exaggerated the ferociousness of these peaceful, gentle animals.

E.P. Since we arrived at its enclosure, it interests me that this chimpanzee has tried to clap several times and draw our attention, as though he wants to communicate. . . .

J.S.P. He tries to communicate even though he doesn't know how to, and we. . . .

E.P. Less so.

J.S.P. Even less so. The chimpanzee is making intelligible gestures, but we hardly know how to do anything.

E.P. Since we were very young, we have been told that we are totally different from animals. One of the biggest arguments was that, unlike human beings, animals could not talk. Jordi, how do you respond to such a prejudice?

J.S.P. Although it is said constantly, it is not true. A barrier has been erected, a wide chasm almost impossible to cross. It is true that other animals cannot talk, but they possess a complex communication system. In captivity, they can use up to five hundred different words.

E.P. Between them?

J.S.P. With us. We don't know if it's between them, because our efforts are always directed at them to learn our language, even though we don't know much about their communication. Indeed, in this sense there is a great deal of inequality. Under experimental conditions we can teach chimpanzees five hundred words, and they manage to use personal pronouns. They are also aware of their own system.

E.P. Another of the arguments is that we are different from animals because they don't possess any self-awareness.

J.S.P. They are aware; they do have a concept of their own being. They know they are different from others.

E.P. And chimpanzees recognize themselves in the mirror, right?

J.S.P. They recognize themselves in front of the mirror, and they even demonstrate notions of "me": I am me and you are you. We are two different beings. Until recently this was totally unknown, and yet it is very important, because it means that in the evolutionary process of the brain, a transcendent step has been completed: self-recognition. Chimpanzees, for example, are aware of death and loss. When a chimpanzee disappears, the others become very sad. When a mother loses her child, she enters a state of depression that may last for months.

E.P. Let me tell you about something that is hard not to define as shared higher intelligence, as would be described by Nicholas Mackintosh [see chapter 1]. I read that when the dry season hits Africa and there is no water, the chimpanzees, having learned not to drink from putrefied bodies of water to avoid infectious disease, dig and perforate the sides of their dugout wells to get clean drinking water!

J.S.P. Yes. We observed that in a study carried out in Senegal, in the semidesert region of Niokolo-Koba, an extremely arid savanna. The chimpanzees would not dare drink from stagnant putrefied pools. Rather they dug wells and drank fresh water.

E.P. How is this acquired experience, or mastered innovative skill, transmitted to the others?

J.S.P. It is transmitted culturally. Young chimpanzees and other chimpanzees observe this, they learn it, and it starts to form part of their cultural tradition.

E.P. Instead of turning animals into a zoological show, have we tried to teach them something? Couldn't they learn to sweep, for example?

J.S.P. It would make no sense to them. They are not made for such things. They follow the evolutionary line of forests, savannas, steppes, jungles, et cetera. They say that if we made gorillas and chimpanzees carry out human tasks, they would eventually turn into human beings. At any rate, it would be absurd to try to humanize them, because they will never be human. They diverged; they are heading in a different direction, and there is no reason why they should sweep.

E.P. Look at that chimpanzee. He has been using a stick, introducing it into a rock with the hope of finding something underneath. . . .

J.S.P. A reward.

E.P. A reward? He is then taking the stick out and putting it in his mouth, right?

J.S.P. Yes. Thirty years ago we discovered that chimpanzees make simple tools, like sticks of a certain size to get termites. If you look at these rocks, they are an imitation of the termite mounds of the savanna, but instead of termites, we put a bit of honey there. The discovery showed that chimpanzees are capable, in a cultural sense, of learning how to make simple tools from natural objects. This is cultural behavior.

E.P. I read in one of your writings that you believe humans are not only not very different from these animals, but that we started in a very similar way. You also said that even though it seems absurd, we used to be tree dwellers like many other animals.

J.S.P. I am convinced that until fire was discovered, humans lived in trees, like chimpanzees do today. Why? Because they lacked night vision. The human retina is not very precise, and it cannot capture crepuscular light in the same way as night birds of prey. Therefore, hominoids could not sleep on the ground for fear of being eaten by predators, and so they sought refuge in the trees. Half a million years ago fire was discovered, freeing the hominoids from this form of "imprisonment," for lack of a better word, that was having to sleep in the treetop nests of the chimpanzees.

E.P. Think about the future. What does it hold for our bizarre relationship with the other animals?

J.S.P. I find our relationship with animals nothing short of tragic. The human pursuit of knowledge and experimentation has its roots in religion, because animals are seen as subservient to us, which is not true. Plants and animals are our brothers and sisters. We share a great deal of DNA in common. Human DNA is very similar to plant DNA, and it is especially similar to that of mammals.

E.P. You believe that we will be judged in the near future for our ruthless treatment of animals.

J.S.P. Future generations will judge us in the same way we judge the slave traders of a hundred years ago. Slavery was not abolished in Spain until one hundred years ago.

E.P. And you believe that the day will come when we will abolish—

J.S.P. I believe that humanity is on the road to abolishing the mistreatment of animals, but when it finally makes it into legislation, there will not be any animals left in the wild to protect.

— 5 —

Like Ants, Bees, and Termites, but Different

Interview with Edward O. Wilson

We are the destructive meteorite.
We never fully give up our individual interests;
that is why we are not a superorganism like ants.
—Edward O. Wilson

Edward O. Wilson has his study on the most beautiful floor of Harvard University's Museum of Comparative Zoology. Surrounded by photos of anthills—the study of ants is the subject to which he has dedicated much of his academic life—he has concluded, to him irrefutably, that only one attribute

---

Edward O. Wilson is the Pellegrino Research Professor in Entomology (Emeritus) in the Department of Organismic and Evolutionary Biology at Harvard University.

separates humans from superorganisms like those of ants, wasps, or termites: hominids never totally give up the defense of their individuality for the interest of the hive or, in our case, the community.

Eduardo Punset: For thousands of millions of years, life has been more or less predictable, so much so that you scientists have developed models of life. You yourself said that there could be bacteria anywhere on Earth and that if there were water in which to swim, small protists would immediately appear and, like predators, invade it. And, by analysis of evolution, we can predict laws. Take living space: the larger its volume, the larger is the animal. The more stable the climate, the more species there are in the region. In the next one thousand years, will we be able to predict life and its models in the same way?

Edward O. Wilson: The great image of the evolution of life on Earth has been one of expansion: from the start, with very small, simple, single-cell organisms, over 3.5 billion years ago, to the present day; from the marine waters to the land, freshwater and the air. So, today life resides in every conceivable place: Wherever there is water, or the promise of water, one sees life—at least in the form of bacteria, microscopic protists, or fungi. Life is present from pole to pole, from the Antarctic to the icecap of the North Pole, from the summit of Mount Everest to the Challenger Deep, twelve thousand meters beneath the sea surface. This is one great model: life encompasses the entire planet, in every conceivable habitat. And another important model in the evolution of life over this immense period of time is regular expansion and diversity of life forms. From a small number of species—if bacteria can be claimed to have species—to today when the number of species is unknown. We know that there are between 1.5 and 1.8 million identified and described species with scientific names, but the estimate of the number of species on Earth varies between 3.5 million and 100 million! No one knows for sure because so little is known about microbes (bacteria, protists, and fungi) in nature, the smallest organisms among us.

E.P. And this diversification of the species is not easy to understand, as we all come from just a few. So, from such uniformity, how on earth have we arrived at such diversity? What are the motives for the diversification of the species? What made them create themselves?

E.O.W. We understand factors are determined by the appearance of certain species in one place in particular on Earth. The easiest way to remember them is with the letters ESA.* E is energy first of all. The more energy there is—especially solar energy—the more species can be maintained. More energy, more species. This is the reason there are more species in the equatorial regions, especially the tropical, most humid areas of the tropical jungle, the rainforests. The S stands for stability. The longer an area has remained the same, that is, the number of millions of years that an area has remained unchanged, the greater the diversity of species. As evolution continues, the species adjust well not only to the environment but also to each other by means of symbiosis. They combine together in systems with greater stablility. This is the reason for the stability factor. The seabed at a depth of thousands of meters and in complete darkness, for example, is inhabited by a large number of species despite the lack of energy, because conditions have remained the same for millions of years.† And the A refers to area, really to volume. The larger the area, the more different species can thrive sustainably. So, on a very small island in the West Indies, of seven square kilometers, there may be no more than three or four types of lizards or one reptile species. A medium-sized island like Jamaica or Puerto Rico may have twenty or thirty, whereas an island as large as Cuba might have a hundred. This is an important factor: the larger the area in which the species can live, the greater the number of species can be found.

E.P. So the time has come when, despite large areas and available energy, we have reached a bottleneck, in the sense that there have been huge population increases and much greater consumption of energy. As you said, if the rest of the world reached the level of consumption of the United States, its technology, we would not require one Earth, but more than five planets like ours! Why? To supply industrial resources, food, and freshwater—things called "externalities" by economists!

E.O.W. During the past 450 million years, major extinctions have occurred. We know one very well: the Cretaceous period of the Mesozoic era, one that, due to a gigantic meteorite, happened at the end of the dinosaurs'

*Not to be confused with the much better-known abbreviation of Europe's equivalent to NASA, the European Space Agency (ESA).—LM

†Although no sunlight reaches the seabed chemical energy, bacterial oxidation of gases such as methane (CH_4), hydrogen sulfide (H_2S), ammonia (NH_3), and hydrogen (H_2) supplies energy for the underwater ecosystem. In localities such as the deep sea vents and the cold seeps off the Florida Coast the animal life of the community depends on an energy supply: chemical oxidation of specific inorganic components (Stewart, 1999).—LM

reign, 65 million years ago. But an even larger extinction occurred during the Permian period, more than 200 million years ago. Possibly it too was due to meteoric impact. So, life continually expands. Then much of it is extinguished. These dramatic changes are detected every hundred million years or so. Afterwards some 10 million years are required to restore diversity and evolution continues. One of the great extinctions, the one that took place 65 million years ago, was preceded by an enormous expansion of life that produced what we now know as the mammals, including us; the insects, which expanded enormously; and the plants with flowers that replaced the conifer trees like the pines and firs.

E.P. You say it takes 10 million years to recover?

E.O.W. Yes, approximately 10 million years, after a period of great extinction. What I'm saying is that probably it took between 10 and 20 thousand years before human beings expanded to the far corners of the planet. An agricultural revolution permitted the consolidation of the far-flung population. Possibly here in the Holocene, or just before 10 or 20 thousand years ago, life hit a peak of diversity. Then we appeared. We are the great meteorite.

E.P. We are the meteorite?

E.O.W. Yes, we are. Right now, human activity reduces diversity and we are facing the first stages of the "sixth extinction." This is the "bottleneck" about which so much has been written. The bottleneck is human overpopulation. Humans tend to destroy so much of the natural environment that other species are no longer able to maintain themselves. Also it is related to increase in per capita consumption as people, all around the world, increase in the quantity of foods and resources consumed. The combination of an increase in population and personal consumption leads to the elimination of what could be called the "natural capital" of the world. We inherited a certain natural capital, an inheritance of an economic nature that forms part of a market economy that we now use and destroy. The bottleneck idea has an element of hope based on the fact that although there are now over six billion people in the world, the rate of population growth is falling. It seems that women, at a certain level of education and independence, choose to determine their own lives and their number of children.

E.P. They decide to have fewer. . . .

E.O.W. They opt for quality and decide to have just a few children, instead of a great number scattered around, like the seeds of trees. And women throughout the world, and especially those in the industrialized nations of Europe, the United States, and now Asia, reduce their fertility levels and number of births to below what we call "breaking point," slightly more than two children per woman. When we reach this we have zero growth in population. The United Nations predicts that we will reach a maximum of nine billion people, which is half again the current number.

E.P. And only then will we be able to overcome the bottleneck?

E.O.W. But the problem is how to overcome it without destruction of planet Earth and without destruction of so much of the rest of life, which is what we do.

E.P. You mention that our brain is related to the bottleneck. You said, in fact, both things: there is a surprising indifference of the brain toward the environment and toward the protection of species. Our brains are concerned with very small geographical areas. We are too worried by just close family. We have a very narrow perspective, of one, two, or three—no more than five—generations. But on the other hand, our brain shows signs of biophilia? . . . Our innate affinity with nature, is that right?

E.O.W. Much of the difficulty of the bottleneck comes from the great level of consumerism in the industrialized world. The United States leads this consumption. As you mentioned, for everyone in the world—about six billion people—to live at the same level of consumption as the United States, four or five other Earthlike planets would be needed. This poses tremendous problems if we want to maintain or improve our quality of life, while we reduce consumption. This great technological challenge is currently faced by mankind. The worldwide tendency to excessively consume, aggressively and competitively, is a manifestation of human nature. It is related to the fact that our brains mandate aggression, competition, pride. There's no doubt that we are influenced by our biological nature. One feature of human nature is the tendency for women to reduce the fertility rate when they are independent. This is a marvelous, providential quality of human nature. Another is to think only in the short term. At the very most, when we think of the future, we think of the next generation and only in a very small space, at most—in our own community or perhaps our own country. We make terrible mistakes in the fields of economic and resource planning. It is very easy to start a war and adopt aggressive behavior due to this lack of vision. We must overcome these problems. The best way is by understanding ourselves. The current approach is the subproduct of the past: Aggression was a beneficial type of behavior when humanity lived in small groups and its tribes were scattered around the world. From a Darwinian point of view, survival and reproduction were imperative, so aggression and competitive behavior were very intelligent. If we only think in the short term, we have to do things well and survive until tomorrow. We must face the enemies that surround us. Now that we have more knowledge, we can take into account the worldwide problems we should overcome. We can also overcome our tendency to destroy other organisms and the biosphere, i.e., they that keep us alive. We realize that we are all able to behave like Hitlers. Even if we don't care about the future human generations when we destroy other life, we lose something very important to the human spirit, for the soul—if we can use this word. From everything we know about mankind's evolution and that of all animals, it is logical to expect that, if as human beings we evolved in a particular habitat, on the savanna or in the tropical forest, our brains also acquired an instinct of preference for that environment.

E.P. That's what you call biophilia.

E.O.W. Yes, biophilia, that's right. A lot of evidence favors the idea of biophilia: our tendency to head toward and enjoy particular environments and the other life forms in them. We have come a long way to this "civilization" and must go further until we are able to enjoy our traditional natural environments, and have them at our disposal, even to go there and live.

E.P. There is also, we hope, this pride or pleasure that we share the same genes with so many species. I think you've written that the "rest of life" is the body and we are the "mind." We are here to protect the rest of the body. Are you confident, after fighting for so many years? Your first major book already said that behavior is based on biology and nature, and that was thirty years ago. . . .

E.O.W. Almost thirty years . . .

E.P. And are you confident that this biophilia, this pleasure in sharing common genes, will bring a cultural change in human beings?

E.O.W. I think so. I think that the understanding of the origin of humans and human nature, as part of our research into how the mind functions, will reveal that we have certain profound psychological needs. Not only do we develop certain relations among ourselves, but also with the other living beings, the rest of life on Earth.

E.P. Another idea found in your way of thinking—and that I find very interesting—is when you try to explain, or remind us, that the conservation of the environment is not contrary to the laws of the economy. You say: "We must realize what we are sacrificing—that which is now free—the enrichment of the soil, the regulation of the climate," and what else?

E.O.W. Or the very air that we breathe and which the rest of life offers to us. In fact, many years ago a team of economists and biologists tried to estimate the value in dollars of the natural world that we are destroying: the water, the air, the soil, et cetera. And the result they arrived at was thirty-three trillion dollars a year!

E.P. Trillion!

E.O.W. Trillion with a "T": yes, a thousand billion. It's the same as the world GDP; in other words, the human production, in economic terms, of the whole world. And it is given to us completely free, and when we destroy the natural world we have to replace it with our own economic machine; in other words, when we destroy a forest or a water reserve that is provided.

E.P. Free of charge . . .

E.O.W. Free water that we destroy, which is something that happens all over the world, then we have to replace it with filtering devices and that costs millions, hundreds of millions of dollars. Step by step, what we are doing is to turn the Earth—literally—into a spaceship in which we as a species cannot relax, we can't sit down and let nature supply us with all those services. We have to be as if we were living in space, in a space vehicle, always repairing, measuring, discussing in order to be able to make things function. It's crazy.

E.P. Automatic well-being has been destroyed and any future well-being will have to be achieved with constant management, including our own reproduction. If extinction is happening worldwide, where should conservation efforts be concentrated? Let's be modest in this regard: in the short term, culture in general will not change sufficiently for a rapid enough significant turnaround. How most effectively should we concentrate our conservation efforts in the near future?

E.O.W. We know exactly what should be done. I've been affiliated with several worldwide nature conservation organizations, for example, the World Wildlife Fund or The Nature Conservancy, for years and all we do is think about this. Each one of these organizations has a budget of about a hundred million dollars a year and they are becoming efficient, though much more is needed. With the money we get we try to save as much biodiversity, as many species, as possible. We attempt to prevent extinction. These species usually are over a million years old, or even more, but we cannot replace them. If they become extinct, we'll never see them again, they will be lost for the human mind. The best way to concentrate our efforts is by direction toward developing countries, in particular those with humid or tropical jungles, because over half of the plant and animal species on Earth are found in the tropical rain forests, although they cover only 6 percent of the surface of the planet.

E.P. Don't they contain 40 percent of all species?

E.O.W. Places called hot spots, inside the larger areas, often the humid tropics, but not always the tropical jungle, cover only a little more than 1 percent of the surface of the planet. Think about it: 1.4 percent of the surface area of the Earth supports 40 percent of all species! If we know exactly where these places are and save them, then we will be able to preserve 40 percent or more of all plants and animals on Earth. If we add the central areas of the rest of the uncultivated land, which are enormous portions of tropical jungle, savanna, the Amazon and its basin, the Congo basin in Africa, and New Guinea, if we add such central areas to the approximately twenty-five hot spots that cover 1.4 percent of the Earth's surface—we have to think in these terms—the estimated cost would be twenty-five billion dollars to preserve the total area of hot spots and 40 percent of the species. It can be done. We should do it!

E.P. And about superorganisms, what are the differences, if they exist, between your idea about superorganisms—you say that we humans are part of a superorganism—and the concept of Gaia as a planet that is also a self-regulating superorganism?

E.O.W. The human species is not a superorganism, nor should human society be called a superorganism. Ours is a mammalian society developed thanks to intelligence that led to reading and writing, to the stipulations of contracts and constitutions. We have the ability to cooperate, even to preserve and improve our personal interests. But without bureaucratic hierarchies, the social ants do decide the best option for the group.

E.P. And our individuality?

E.O.W. I have just finished a book about the superorganism and I have had more time to analyze it. I know what a superorganism really is and what humanity is. A colony of ants is a superorganism. They are incredible, particularly some species. They are marvelous creatures. They have a series of very complex behaviors by means of which the members of the community cooperate. The colonies use and recombine between ten and twenty types of chemical signals.

E.P. Pheromones?

E.O.W. Pheromones, which the members of the colony smell and taste, and they may have up to fifty messages from combinations of pheromones. They have multiple forms of individuals that form groups that perform a single function very well and others not so well. All behave as a united whole, so that for generation after generation the colonies of ants of certain species are always the same. An ant's brain is programmed, almost entirely, to carry out certain types of communication and to work in a very specific way. The individuals are categorized by the function of their work so the colony can survive: the unit is the colony. Evolution continues because of competition between one colony and another. One of the results of this is that there is a certain degree of harmony and cooperation among the individuals in a colony. But the colonies are always at war with each other. Ants are the most warlike creatures on Earth. I say that if we gave ants nuclear weapons they would blow the world up within the week. Let's compare ants with humans. . . .

E.P. What about human beings?

E.O.W. Ants began to exist during the late Cretaceous era (65 to 138 million years ago).* This was the time of the dinosaurs. We have found fossils not only of the first ants, but also of the ants of fifty or sixty million years ago. This is a long time, yet these ants are exactly the same. There had not been much change. Although there was considerable evolution at first, the ants have not changed in the past fifty million years. The same codes, the same models . . .

E.P. Perhaps that is because they are efficient.

E.O.W. Behavior has become much more complicated. Let's analyze human beings. If we consider our most immediate ancestors, the hominids—who already stood upright and used tools, an advance compared to the monkeys—we have only been on the Earth some 5 million years. That is very little time. And nonetheless, once we reach the stage of *Homo sapiens*, the modern human species, the evolution of the brain is, among organ evolution studied, one of the fastest of all time. So, something happened that turned these primates into what we now recognize as human. And in the process we didn't become beings like the ants, but rather continued to be independent mammals: each human being works for his or her own interest. And this individuality and creativity is conserved. So human society, unlike that of the ants, is partly based on instinct. Yes, we have a lot of instinct: human nature determines to a great extent our emotions and what seems satisfactory to us. Based on these orientations, we have created our societies and civilizations through long-term contracts, agreements we consider sacred. Among us, we have arranged agreements, relations, contracts . . . and that is how we do it, but . . .

---

*See Table 1 on page 5.

E.P. Excuse me for interrupting you, but have we inherited this from the social primates? That is, do the social primates make agreements similar to ours?

E.O.W. Yes, in a rudimentary way. Particularly the large apes, the social ones: the chimpanzees and gorillas. And most especially in the gorillas and in a kind of very advanced chimpanzee, the bonobos. They have the rudiments of what we recognize as a human form of contract and agreement, coalition, even conscience. They have the feeling of guilt and . . .

E.P. Punishment—

E.O.W. So we have created a civilization based on that. Even at the current level of evolution, although we continue to act like a colony of ants and make war, our ability to form a social order and very rapidly advance in knowledge and technology within society gives us great hopes of eliminating the worst characteristics of human colonies—

E.P. Ant colonies.

E.O.W. Yes, to get rid of the last vestiges of our behavior that is similar to that of ants. That is the difference. But, obviously, we have to understand the poor ants. On any given day between one and ten quadrillion ants are alive! And consider, moreover, that an ant weighs one-millionth of what a human does. A million ants weigh the same as a human. So all the ants in the world weigh approximately the same as all the humans. The brain of an ant is just one-millionth the size of a human brain. We must admire ants, especially if we analyze what they have managed to achieve with their tiny brains. If you extract it and analyze it, it can't even be seen; their brains resemble a speck of dust. They have achieved incredible things, and with very little change in the last fifty million years.

E.P. You have written something that arouses my curiosity so much that I will quote it. I would like you to tell me what exactly it means: "Only when the mechanics of the brain can be put down on paper, as we do with the cell, and we can reconstruct it based on what we have on paper, will the properties of the emotions and ethical judgment be clarified." What do you mean? What do you hope?

E.O.W. There are two levels of science. One is the precise description of behavior and thought, but at the next level of science we have to take this incredibly complicated series of phenomena and try to reduce them to their basic elements and processes, and then try to understand how they fit together to form the unit. Each science is a constant cycle of reduction and synthesis. And a mature science involves the process of reduction and at least a little synthesis to follow it. The true scientist believes—though it is perhaps not shared by many academics in the humanities or the theologians—that the mind is a product of the brain, that they are not separate and they are physical. Mind-brain is not nonphysical. We should be able to understand the mind in the same way we understand any other methods of the natural sciences.

E.P. With regard to the possibility of intervention in the brain as we already have intervened in the cell, of penetration into the internal potential of the brain, is it possible to fuse the machine with the brain? You know we now enter the age of biological control. In this way we could change our environment and avoid the bottleneck.

E.O.W. It is popular among scientists and those who think of the technological future that we will be able to equalize and unite the brain with the greatest advances in computations. We should simulate feelings and emotions by means of a machine. We like to think that one day a superintelligent intelligence will be developed—without losing individuality—which will be a human brain in combination with very high computational capability. This might happen in the future. I don't think it will destroy our humanity if we manage it correctly. So, will this superintelligence be able to save us and help us solve problems faster? I think so, insofar as it will enable us to understand the world in a more complete way, including the origin of mankind and human relations. Processing knowledge faster must help us to reason, which we do already with the Web: we put all the information in the same place. If we do it better and better, there is the hope that individual wisdom will be acquired more quickly, and from there a collective wisdom.

E.P. In your most recent research, you highlight the fact that human beings, unlike ants, cling to an individuality that defines them as humans and differentiates us from superorganisms. It's fantastic. May I ask you a question to which no one has so far been able to give me an answer? What on earth happened? You mentioned that there has been nothing like it in the evolution of species. What led us to diverge from the ants and make the leap from their type of intelligence to that of *Homo sapiens*? Do we now have a better understanding of what happened than we did a few years ago? Some people say that it is possible it was a change of diet, the consumption of fish. Do we know?

E.O.W. First of all, the origin of human beings is unique. It took thousands of millions of years to create a species like mankind. That step forward is one of the great mysteries of biology. But it is not the only example of exceptionality in evolution. There have been many cases in the history of evolution in which a great advance has occurred. For example, in my research we come across the work of the *Acromyrmex octospinosus* ants, from tropical America. They are gardeners and build immense underground gardens. They chew and process the leaves on which they grow the fungi on which they feed. They have become leaders of the animals that eat plants in tropical America. But this happened only once in the history of evolution. How do such major events happen? What are the particular circumstances that led to the evolution of humans? We don't have the answer, but we will continue to try to investigate one of the most important challenges in future biological research.

Attractiveness

INTRODUCTION

Early in the Proterozoic eon (see page 5) animal ancestors evolved from choanomastigotes cells. Fertilization of egg by sperm became an obligatory prerequisite for maturation. Of course the name we give this fact is sexual reproduction. But don't be misled: sex is not required or even correlated with reproduction in most of the diverse living world. Since, in animals, attraction of egg and sperm is required for evolutionary continuity, attractiveness became of utmost concern to our ancestors long before the fertilization event even produced an embryo, a stage in development that would be visible to the unaided eye. This is why we find so fascinating Punset's interviews here about beauty, choice of mates, and the emotions that stimulate successful mating behavior. Scientists, although extremely limited in their inquiries, tend to be more responsible, more grounded and trustworthy about their admittedly limited conclusions than they are about any facts beyond their speciality. A single molecule in extraordinarily low concentration can induce fatal attraction. Much remains to be learned about the chemistry and behavior of attraction in the thirty million extant animals that perform sex acts as precursors to embryo development.

— 6 —

Can Beauty Be Measured?

Interview with Victor Johnston

> If we measure different parts of the body, we can see just how asymmetrical a person is. This is a highly sensitive indicator of the immunological system in all species. The fewer the asymmetries, the better is the immunological system. That is why we all prefer more symmetrical people, we find them more beautiful.
>
> —VICTOR JOHNSTON

In the past fifteen years, which is not long from the point of view of the geological time repeatedly alluded to in this book, but is significant compared to life expectancy, I have been immersed in the technological and scientific world. No one will be surprised if I say that the results of scientific inquiry, always

---

Victor Johnston is Professor Emeritus of Psychology at New Mexico State University, Las Cruces.

esoteric and arcane, have not yet had much impact on popular culture. People's daily lives remain unaffected by the unintelligible deliberations of the scientific research councils.

It is, however, strange to discover that something similar is also true in the scientific community, but in reverse: Many scientists would consider it frivolous to think that scientific discoveries have anything to do with people's daily lives. For that majority of scientists, the aim of science is not to solve the problems of daily life, but rather to gain a deeper understanding of basic knowledge that can and should lead nowhere.

That idea is a gross error that has contributed to cementing the barrier between science and society, a barrier that will crumble and fall in coming years and which the geologists and paleontologists have been the first to break through. It is they who have excavated rocks and fossils from hundreds of millions of years ago—could there be anything more apparently remote from the immediate interests of the citizens of a modern city?—and have situated humanity in its geological perspective. They have redefined nothing less than the human condition and its evolution. They have told the man in the street, lost in the midst of the deafening traffic, exactly where he stands. Could anyone contribute a comparable tangible asset?

And, as often happens in the history of science and of life, the most important person is not the best known. In this specific case, not even many scientists know the founder of their discipline. In the seventeenth century, the century of the scientific revolution, both Protestant and Catholic theologians—so starkly opposed on other questions—shared the conviction that the only source of primordial information on the Earth was the Book of Genesis. The scientists themselves were perfecting "experimental philosophy" and they all thought it absurd that there could be a science of the remote past. The merit of the demonstration that from an inert rock, from fossils, one could extract an understanding of living organisms must go to Nicolai Stenonis—or Nicolaus Steno (1638–86) in the less florid, more common version of his name. A Dane who worked in the Netherlands, France, and Italy, he was the first anatomist, geologist, paleontologist, and, more generally, scientist of the scientific revolution.

Moreover, it is very difficult to identify a field of basic research that has not had practical consequences: the equation $E = MC^2$, formulated by Einstein in 1905, made it possible to discover that inside the Sun, mass is transformed into energy, which in turn energizes life on Earth. The Milankovitch cycles of the Earth's orbit have helped us understand the measurements of the level of CO_2

in the atmosphere and to demonstrate that the concentration of CO_2 now is higher than it has been in the past three hundred million years. Dirac's reflections on antimatter led to positron emission tomography (PET) techniques. In formulating of the new geometry of space-time, Einstein discovered that the rhythm of clocks depends on gravity and that therefore the people on the upper floors age more slowly than those on the ground floor. The sequencing of the human genome, an abstract task if ever there was one, has already made possible the cure of diseases—very few—due to a single gene mutation.

The reader will therefore not be at all surprised that we begin this series of conversations with scientists by dealing with the application of science to one of the most familiar subjects for the man in the street: the passion for beauty. The fact is, when two people fall in love as they catch a glance of each other out of their respective car windows one Monday morning, it is not a cultural whim or a sign of sexism, but rather the manifestation of a basic instinct that has its own markers inside the brain. We pay tribute here to Steno, who reminds us that neither fossils nor sexual seductions escape accessibility to scientific investigation. In later chapters we talk about the universe and nanotechnology.

Probably the best person to speak about how we perceive beauty and why some faces are attractive to everyone is Victor Johnston, biopsychologist at the University of New Mexico.

Victor Johnston: It forms part of our basic nature: generally, women and men are attracted to the opposite sex, and this has been imprinted on our brain since the very start of life.

Eduardo Punset: How does it work? What kind of signs do we look for?

V.J. We look for attractive characteristics in the opposite gender. Men like faces that show low levels of testosterone, the male sex hormone. The face of an attractive female has a short lower jaw, which is the sign of a low level of testosterone, and fleshy lips, an indicator of estrogen, the female sexual hormone. A high level of estrogen and a low level of testosterone indicate fertility. A face with these hormonal markers that indicate a high level of fertility is attractive to men.

E.P. Like the proportion of 0.66 between the waist and the hips, another sign of fertility that coincides with that of the Venus de Milo?

V.J. Yes.

E.P. Why is long, healthy hair another indicator of fertility?

V.J. The word "healthy" is the key. Health is essential to fertility. A person must be healthy to reproduce well; he or she must have attractive skin and healthy hair. These characteristics are also very attractive for people of both sexes.

E.P. Victor, tell me something. When do we start to perceive the attraction of beauty in the brain receptors?

V.J. This is very difficult to establish. We know that the hormones that influence our brains at a very early stage of our development, mainly testosterone, change the structure of the brain and subsequently make us more sensitive to hormonal markers. So, an important question concerns the influence of hormones on the brain in the uterus, very early on. Perhaps the brain is subjected to hormone influence by the thirteenth week of embryonic life, the time at which the sex of the fetal brain is determined. And already that influences what the person will find attractive in later life. Men's brains will feel attracted mainly by women's faces, and vice versa. Hormones not only influence the development of the brain, but also that of the body.

E.P. But it would appear, according to your research and that of your colleagues, that very early, by the age of two months, a baby is already attracted by one thing or another?

V.J. Yes. From the time they are very young, babies are attracted by normal, run-of-the-mill faces. But as they grow and reach adolescence, they change and start to feel attracted by faces with more marked features. Girls are attracted to faces with more masculine features, and vice versa for men.

E.P. You are suggesting that by means of natural selection, people feel an attraction toward the average face at first, but when sexual competition begins we forget about the average and prefer the girl that is more feminine?

V.J. Exactly. It is natural selection to draw attention to yourself by being more attractive than the average, like the peacock who displays his tail, even though that tail is not at all useful for its survival. It is not a product of natural selection, but of sexual selection and it does serve to attract the opposite sex, the peahen. The tail is a very important indicator of the male health for the female.

E.P. Even though it's a burden.

V.J. Yes, the tail is a burden and correlates with a certain reproductive loss, as the longevity is reduced: the peacock doesn't live so long because of his long tail, and it's easier for predators to hunt him. But those reproductive losses are more than offset by the advantages of attractivity to the female, so the tail is a product of sexual selection. The same thing happens with the physiognomic characteristics of men's and women's faces. They attract the opposite sex because they say something very important. In the case of women, they say "fertility," and we think that in men they say "healthy immunological system." We think it's important for women to access men's immunological system, because they will mix his genes with theirs. If the man has good genes, his children will survive and, of course, his genes will also survive. So what women look for is some symbol of a man's immunological competence, of his good genes.

E.P. So even now these systems, which would seem to be very reasonable sixty thousand years ago, are still valid. It's surprising to me that your research tells us that still today it is beneficial to be good looking.

V.J. Of course.

E.P. Has that been demonstrated?

V.J. Studies demonstrate that beauty affects how easily one finds work, or the decisions people make about us. There's measurable leniency toward attractive people. In courts of law more attractive people receive shorter prison sentences or are declared innocent more frequently. Beauty also has a great influence in recruitment; good-looking people more easily find work. It's not fair, but it's the truth.

In the hunter-gatherer society, if you liked sugar, you ate mature fruit and enjoyed a very healthy diet. Today, in sugar refineries sugar is separated from food that contains it. We still seek sugar, even though now it kills us. Something similar occurs with beauty: We still seek it, even though we don't need it, because drugs can increase fertility, contraceptives, and many other interventions unrelated to beauty can manipulate fertility.

E.P. But obsession and basic instincts—

V.J. Basic instincts haven't been lost and apparently they will continue to exist for a long time.

E.P. This is almost irrelevant, but I was very surprised to learn that beauty has its own territory. In other words, when a tall and good-looking man approaches another person of similar stature, he will remain at a distance of half a meter. But if the good-looking person approaches someone smaller, that same distance will not be respected. It is as if the good-looking tall person carried his own larger territory, personal space, with him, relative to the smaller man.

V.J. Good-looking people appear to be more intelligent. They dominate more. It's strange, but this is the perception in our society. We give way to good-looking people, we open doors for them, we give them more space, we are more indulgent, and allow them to make more mistakes.

E.P. Have you investigated whether intelligent women are equally likely to marry as attractive ones?

V.J. No, we've mainly concentrated on the physical characteristics that make people attractive. We try to understand why certain characteristics are attractive because we believe there is something biologically important here.

E.P. Let me ask you this: we knew, somehow, that a woman's sexual activity is regulated by her menstrual cycle; but you say that not only is this true, but women will choose one man's face or another depending on something that you call the "digit ratio."

V.J. Yes, that's right. Recently we have discovered that women change their preferences during the period of the menstrual cycle in which there is a greater risk of pregnancy, just prior to ovulation. In what direction does the change take place? It depends to what extent the brain has been affected by the testosterone present in the uterus.

E.P. As a fetus?

V.J. Around the thirteenth week of embryological life, the increase in testosterone effect becomes detectable in the fetus. It is determined by measuring their fingers. The greater quantity of testosterone to which a male fetus is exposed in utero, the more the ring finger grows. And that makes him, in a way, less feminine. When adult, a woman with a long ring finger feels attracted, when she experiences a high risk of pregnancy, by more masculine men: men with prominent, square jaws. She simply searches for men with "good genes." More feminine women, with clearly shorter ring fingers, tend in the opposite direction. When she ovulates, thus has a high risk of pregnancy, she seeks friendlier, gentler men.

E.P. You say that the digit ratio in women is one or greater than one; that is, the length of the index and ring fingers is equal, or the index finger is longer. And that is determined genetically.

V.J. Not only genetically, but hormonally. The digit ratio depends on the testosterone present in her mother's uterus at a very early age.

E.P. And how does this influence determine the woman's choice of a certain type of face during the menstrual cycle?

V.J. A woman who as a fetus has been exposed to the highest levels of testosterone has a typically masculine digit ratio: a longer ring finger. These women tend to prefer very masculine men. They want masculine men for both sporadic and longer-lasting encounters. Even when there is a high risk

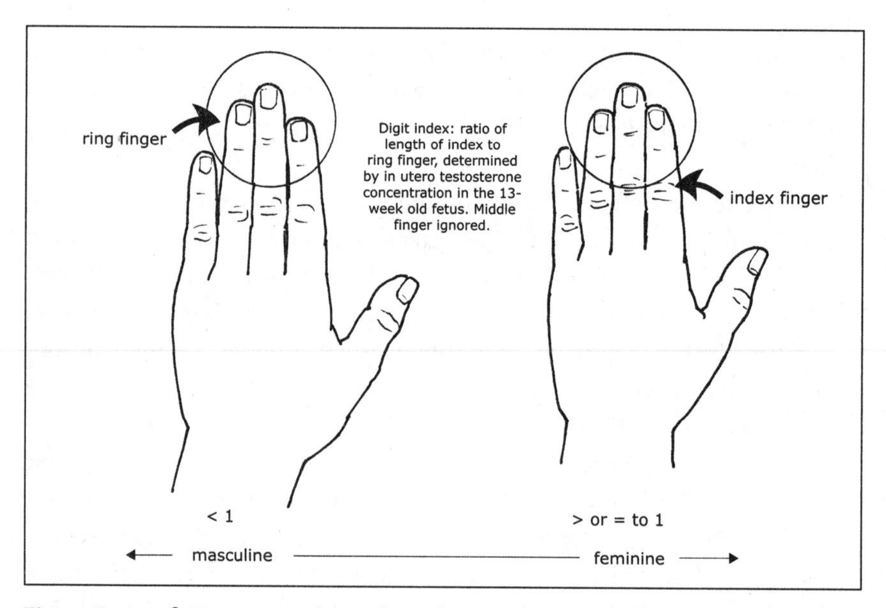

Fig. 1. Ratio of Finger Length resulting from Fetal Exposure to Testosterone. The hands, of both men and women, are testimony to the hormonal flows at the fetal stage. The length of the ring relative to the index finger indicates level of exposure in the fetus to testosterone. Greater exposure to testosterone has occurred in those whose ring finger is longer than the index finger; exposure to less testosterone is reflected in more equal ring and index fingers or a ring finger shorter than the index finger.

of pregnancy they still prefer very masculine men. So these women, who to a certain extent have been "masculinized" by the testosterone in the uterus, show these preferences throughout their lives. On the other hand, women with a more feminine digit ratio, one-to-one or shorter ringer fingers, may prefer masculine men for short relationships, but when it comes to a longer-lasting relationship, or when there is a high risk of pregnancy, they prefer a gentler, more amiable man. They choose not the typical macho man, but rather someone who is closer to average.

E.P. Victor, will this ever give us any keys to sexual behavior? I'm thinking about bisexuals compared to heterosexuals, for example.

V.J. This is a very controversial subject. Many studies demonstrate differences between men's and women's brains as a result of exposure to testosterone in the uterus. The brains of homosexuals do differ; they tend to be midway between a man's brain and a woman's brain. So, we begin to understand how this hormonal context influences our brain structure and, probably, the sexual preferences we show in the course of our lives.

E.P. Some evolutionary psychologists say that the digit ratio is perhaps simply a reflection of natural selection, to the extent that a longer ring finger helps to stabilize the middle finger, so giving you greater control, for example, over stones or arrows. It's a hypothesis.

V.J. More and more evidence shows this digit ratio to be related to factors such as sperm count. In men who have a longer ring finger (a smaller digit ratio), the sperm count is greater, for example. Digit ratio is also related to athletic ability, especially in the case of runners: we could almost predict who will win a race by observing the digit ratio. In a BBC program, an investigator who knew how to interpret the digit ratio was able to say who would win a race before it began. So the growth of this digit ratio is related to many masculine characteristics. Therefore, I think it is much more significant than simply a matter of stabilization for a hand that needs to throw objects: there is something fundamental related to the digit ratio and it is manifested in many different behaviors.

E.P. It's fascinating that beauty is so predetermined. For example—correct me if I'm wrong—people don't generally like short, fat men; they tend to prefer tall, slim men with long legs. Is it true that this is because in the past we were runners in a tropical climate, where there had to be a low skin-to-body-mass ratio?

V.J. I think that is partly true. Testosterone causes growth in puberty, when we become taller. Therefore, height is always an indicator of exposure to high levels of testosterone. We tend to like tall, dark, handsome men: there are physiological reasons that mean they will probably be good companions.

E.P. And color? Dark-skinned or black . . . ?

V.J. In the majority of societies people prefer a lighter skin tone than the average. This may be a useful indicator. We tend to associate dark skin with age and some people prefer a lighter color because they think that in some way this corresponds to a younger person. I don't know.

E.P. And in terms of natural selection, it is said that pale skin helps to absorb sunlight, so one makes more vitamin D. This seems reasonable.

V.J. Yes, that's true. As we move north we find paler skins, probably in order to be able to absorb more sunlight and increase the production of vitamin D.

E.P. What does your investigation suggest?

V.J. I think we demonstrate that hormonal influences change the structure of our brain. The changes take place at a very early stage of life. And there is a fundamental difference between men and women: There is a difference in the structure of the brain as the result of hormonal influences. What is fascinating is that these hormonal changes influence our choice of partners later in life. I think that in our research we try to understand why this happens and to detect certain hormonal clues that influence attractiveness, as a function of physiology that took place in the remote past in the uterus.

E.P. In the uterus. But what will happen, or what is happening, when in reality, through cosmetics, fashion, and plastic surgery, we change these signals that are the end product of exposure to hormones? Our hair, lips, the distance between the eyes? Will these alterations give rise to social changes? Are the normal and social changes compatible?

V.J. Each culture has its prototype of beauty. We have seen that, historically, in all cultures that have been studied, people have tried to change their appearance to appear more attractive. This is nothing new, it is simply that now we can do it faster and better. We can manipulate faces and bodies much more than before, but all the changes still conform to an old pattern that we human beings have followed since history began. We have always manipulated our appearance to appear to be more attractive than we really are! We have been deceiving Mother Nature for a long time. Simply, we no longer need nature to indicate fertility. In the past the face was an indicator of fertility; now the face and body indicators are not so necessary because we can easily manipulate fertility with drugs.

E.P. But we continue to look for signs of fertility.

V.J. Yes, we still feel an attraction, even though this attraction is no longer a necessary signal. I imagine we will continue to change our physical appearance for a long time yet.

E.P. Another aspect that fascinates me is symmetry as a marker of beauty. What does symmetry have to do with fertility? But, nonetheless, it has been demonstrated that we are attracted by symmetry.

V.J. What is important is not symmetry, it is the extent, the quality, of the asymmetry. If we measure different parts of the body on the left and right, such as the arms, the legs, and the face, we see how asymmetrical a person is. Symmetry is a highly sensitive indicator of the immunological system in all species. It's as if we're born with a perfect immune system and, in the course of development, the symmetry is exposed to viruses, bacteria, and parasites. It is a biological plan: If symmetry is lost, it's perhaps a sign that the immunological system is impaired. So we prefer more symmetrical people, because it indicates the quality of the immunological system. And these hormonal markers—for example, in a man's face, a large lower jaw—correspond to low fluctuating asymmetry. In other words, men with large lower jaws are more symmetrical and have good immunological systems. This is one reason why they are attractive.

E.P. Another mystery is almost as profound as that of symmetry: the question of pheromones or perfumes. I have heard that species that have developed color perception forget about perception of pheromones or perfumes. According to this hypothesis, this is the reason why we—who have developed as social primates and therefore with the ability to see colors—do not need to perceive smells. It that true? Have you studied it?

V.J. I think the perfume industry would not agree with you. If we think about it, it's very strange that human beings have lost the hair on most of the body but we have conserved it on the face. When they reach puberty men develop beards and mustaches. Yet, if we observe the face of a man during puberty, acne appears, which consists of large glands that open and create chemical products that cover a very large area of the beard and mustache. This very large facial area is good for the dissemination of odors. When we kiss, we put our mouths together, and the beard and mustache are just under the nose of the person we kiss. Here, we could argue, circumstantially, that we still use these pheromones and that we produce chemical products that influence the opposite sex. This field is still not well understood, but certain chemical products that are segregated in our glands, such as androsterol, are only perceived by women at certain times in their menstrual cycle. There is a certain type of chemical influence. And I must admit that although the evidence is still weak, I don't rule out the possibility that we humans still, like the ants, use pheromones as part of our sexual arousal system.

— 7 —

Science of Happiness

Interview with Daniel Gilbert

> There is a psychological immune system that protects us from the "slings and arrows of outrageous fortune". . . but we are unaware of it.
>
> —Daniel Gilbert

Eduardo Punset: The great American parasitologist Lawrence Ash—I was with him a few months ago. And, you know, after hearing him talk about worms and other parasites, I said, "Listen, you have the most terrible profession in the world," and he laughed. But you, working on happiness, probably you don't have the most fantastic profession in the world? Do you

---

Daniel Gilbert is Professor of Psychology at Harvard University and director of the Social Cognition and Emotion Lab.

really know what happens with us, when we are happy or when we are unhappy? Daniel, can we really study happiness?

Daniel Gilbert: Well, you know, the study of human beings is, of course, more complicated than the study of parasites and other animals, for many reasons; first of all, people are much more complex organisms. People respond to being studied, while parasites do not. Parasites do not care if you are watching them or not; people do. And also, of course, with human beings, especially if you study happiness, you are interested in something inside them that cannot be observed. They have to reveal their feelings to you through their actions and through their words. So, yes, you may dissect the parasite and look inside it, but we cannot open up human beings and see happiness. All these things make happiness very challenging to study, but, of course, also very rewarding.

E.P. What are the things you insist the most? I mean, you are famous for obtaining information about happiness and for seeing, even measuring, "something inside them that can't be observed." This is the something that you call the "impact bias"? The fact that people, when they are trying to forecast affection, or love, or happiness, they always get it wrong. They anticipate more of it, they overestimate it. Why on earth does this happen? Or how?

D.G. Let us be clear about a few things. First of all, it is true that people do make mistakes when trying to predict how happy they will be, and how long they will be happy after some future event. What is interesting about these mistakes is that they are not always at random. You know, when somebody wants to try to throw a dart at the dartboard and they miss the bull's-eye, that is not so interesting; everybody misses it. But if they always miss it by three feet to the left, then something is interesting there for scientists to study. When people try to predict their own happiness in the future, they always miss three feet to the left. The form their error takes is highly specific. And as you said, there is the tendency to overestimate how happy they will be, or how unhappy they will be. So, both of these mistakes are mistakes of overestimation.

E.P. But, Daniel, what is amazing is that they do this: they commit these errors, with a very common sort of ordinary things. I mean, falling in love, or voting—"If my party loses, I will emigrate to another country"—or whatever. That is incredible—I mean—they should know better because they fall in love every day, they vote every year. . . . And, however, this error persists.

D.G. Yes. A remarkable and interesting aspect of human beings is that they often make the same mistake twice. And then three times. And then four times. But skipping an error is not quite as simple as you are suggesting, because to avoid an error, one has to admit that one has made it. One thing we discovered in our research is that people mispredict how happy they will be, but they also misremember how happy they were. People's "prospections," looking forward in anticipation, and the "retrospections," looking backward in recollection, often match very well. It is just that neither of these match the experience that they have. I give you an example: If you ask voters how they are going to feel before an election if their candidate loses, they say: "Oh, I am going to feel terrible, a month after the election!" A month after the election you measure how they really feel, and they are OK. A month later you ask them to remember how they felt a month after the election, they will say, "I felt terrible!" So you can avoid an error—as we say in psychology—because we are "self-erasing," and we do not realize that we erred.

E.P. Listen. I think that you have suggested that in this business of overestimating happiness or forecasting happiness, there is a sort of psychological—it is a metaphor, I know—a sort of psychological immune system which prepares you for the worst or for the best. Is that true? Have you identified this immune system?

D.G. Yes, certainly. Everybody who has thought about happiness, since Aristotle, has remarked upon the fact that human beings want to be happy, they attempt to find happiness; but when they cannot find it, they find a way to manufacture it. So, this observation about human beings is nothing new. We are enormously capable of changing our view of the world in order to make ourselves feel better about the world where we find ourselves. So, when you are voting against a candidate, you think, "He is a crook, he is a robber, he is stupid, he is dishonest. . . ." But when he wins you say, "You know, I do not like him so much, but he is probably one of the best that we can get." We find ways to rationalize, to think about events that allow us to feel a little bit better about them. What we have discovered, and Aristotle did not know, is that people are unaware of their own talents to change the way they see the world. Every one of us is very good at this, but very few of us realize that we are doing it. Very few of us catch ourselves doing it, very few of us will remember that we ever did it. So, it is as if we have an invisible talent, an invisible shield. There is a psychological immune system that protects us from "the slings and arrows of the outrageous fortune"—Shakespeare said—but we are unaware of it. We all make errors in predicting our own emotional actions. In part, this is why we believe that we are going into the future without an ally. In our brains, every one of us has an ally, a friend, a helper. If something bad happens to us in the future, this helper in our brain helps us to be OK with what does happen.

E.P. From what you say, then the brain is not there to make us happy. The brain is there . . . and now I quote your words: "The brain is there to regulate us," as you just said. Is it true that the brain puts us back to the starting point, so that we do not become overexcited, or overdistressed. . . ? Is this the way it acts?

D.G. I think, yes, yours is a fair summary of what scientists know about emotions like happiness. We experience emotions, we all want to be happy all the time. Let us think what our emotions are for, what they do for mammals. Emotions, we know, for example, are regulated by the most primitive parts of our brains, the parts we share with all the other mammals. We even share emotions and parts of the brain with some nonmammals. What are emotions for? What we currently think in evolutionary biology and evolutionary psychology is that emotions are a guidance system. They warn an animal whether or not to approach, or to avoid stimuli from the world.

E.P. Fight or flight?

D.G. Fight or flight. So, emotions are a kind of mental compass that points us in a certain direction. A compass that always pulls north would be useless, no? So our emotions are not always stuck on "unhappy." They need to be a useful guide to the changing state of the world. Some of the things in the world are good, some of them are bad, for you, for me, as an organism—for our success in reproduction, or ability to continue to live. And if one is always happy, if one never returns to a normal state, then the emotions are accurate sensors of what is useful and what is not. We cannot ever stay in a single emotional state; emotions evolve to fluctuate, like the needle of a compass.

E.P. Is happiness—I do not insist—really not triggered by emotion? As are all emotions, happiness is transient. I mean, there is no eternal happiness. Is that what you are saying?

D.G. But there is no eternal unhappiness, either.

E.P. Not either! That is true!

D.G. Yes, this is very important. It is also very important that we recognize of what we talk about here: We talk about what is true for most people, most of the time. People's emotions are enormously complex and there is a distribution in humans. We differ in any dimension you can consider or measure. Somebody who watches your TV program is happy most of the time. Another is extremely unhappy much of the time. So, we must understand that there are exceptions to emotional rules. But I'm talking about most people, most of the time.

E.P. You were talking about these immune systems, a "psychological immune system." And, apparently—please correct me—when one is affected, under conditions of a disaster, like breaking a leg, or any serious problem, the mechanism of happiness and recovery differ from when everyday frustration comes out in response to trivial problems, like, say, when I leave my dishes on the table in the kitchen. So, what you suggest is that one can be unhappy for very serious reasons, and then this autoimmune system comes into play, or one can be unhappy for trivial reasons, and then one does not feel the unhappiness so keenly. This immune system does not come into play, and, in the end, one may be more unhappy than with the broken leg or other serious problem. Is that true?

D.G. Yes. We demonstrated in laboratory experiments the "psychological immune system." I think people do observe it in their everyday life, too. One of the paradoxes of the psychological immune system is that, like the physical immune system, a critical level of trouble, a serious problem, is required before it is triggered. The implication is that when we experience real drama, events that really upset us, hurt us, and put our self-esteem, or happiness, in danger, the psychological immune system is activated. A divorce, the loss of a parent, the loss of a job, are examples of very big events in a person's life. And very quickly after they happen, the psychological immune system is activated. The activation helps the person find happiness again. Little dramas—what we call "annoyances" in English—lack enough power to activate this psychological immune system. So, although they only make us feel a little bit bad, they continue to make us feel a little bit bad. Perhaps a nice way to say this is that people do not rationalize very small dramas, right? But big ones, e.g., when my wife leaves me, activate the system. When it's activated, I say, "She was never right for me, and I am happy without her, anyway."

E.P. Yes, the psychological immune mechanism becomes activated.

D.G. Yes, it invents some new stories that change the way you think about your world and change the way you feel. When you break your shoelace you do not say, "I am better off without it." Instead you say, "Damn, I need a new shoelace, my shoe still falls off all day." So we have demonstrated that minor annoyances can cause more long-lasting distress to people than major problems. A nice metaphor is physical illness. If you break your leg, you must do something about it. Yes, this is a major difficulty so you go to the hospital and you have it set. In six months your leg is fine. But if you just have a little problem with your knee, you think, "I get this as I get older. . . . It is not big enough to see the doctor. . . ." It does not hurt that much, but it hurts forever, incessantly, because you do nothing about it. This is exactly the logic of the psychological immune system. If the drama is not big enough, the psychological immune system is not activated. And unactivated it simply does nothing to cure us of the drama.

E.P. Daniel, there is something also we do not realize, apparently. From what you say, does having more choice mean that one will be happier? What do you mean exactly?

D.G. Once we understand that we are animals with a psychological immune system, we begin to ask questions like: "What sorts of events trigger our immune system? What sorts of happenings inhibit it?" When we enjoy many possible choices the psychological immune system is inhibited. When we are free to change our minds, the psychological immune system tends not to be activated. The psychological immune system—psychologists have known and demonstrated this in experiments and field studies for over fifty years—the psychological immune system works best when we lack choice, when we are stuck with a serious problem. So, before the election people are not happy with the opposite candidate and his party. It is only once the election is over and there is nothing that can be done. . . .

E.P. They are stuck.

D.G. They are stuck. After the election outcome they suddenly find ways to say that he is not so bad, he actually has some excellent qualities, et cetera. I give you an example from experiments to show how rationalization is very easy to demonstrate experimentally.

We taught a group of students how to take and develop photographs. They took a long course on photography from us. At the end of the course they each had made two beautiful eight-by-ten black-and-white pictures that were very important to them. We told the students that they could only have one of these. They only may keep one and that we had to keep the other. To half of the students we told that after they made their choice, if they ever wanted to change their mind it was fine and we would be glad to trade pictures with them. Other students, we told, once they made a choice, it would be irrevocable. We followed up on all the students and found how much those in the first group liked their pictures and how happy they were by the choices they had made. And the results were very clear. The people were much happier who had no choice, when they could not return their picture. Students who had the options were forever on the fence. They said, "Maybe I made the right choice, maybe I made the wrong one." The psychological immune system was activated to help the ones who had no choice to say for sure that they had made the very best decision.

E.P. Fantastic! This reminds me of the sociobiology of emigrants. We usually say that emigrants, in a given moment in history, were cleverer and that they worked harder, but probably they were stuck. They probably emigrated because they had no other choice.

D.G. Exactly. This is very interesting. Let me add something to the conversation, now. You know the experiment. When you ask people: "Which of these two conditions would they like to be in?" they always choose the one with choices. So it is important. We know that one of these situations makes you happier than the other, but people always choose the one that makes them unhappy. There is an important reason why people make commitments like marriage, make commitments like personal relationships. You know, in America is very popular to get married, less so in Europe, I suppose. . . .

E.P. Still, still . . .

D.G. In the sense that marriage does put us into the irrevocable condition, a condition where it is difficult to change course. When you are just dating somebody, every time he or she does something you do not like, you say to yourself: "Maybe I should not be in this relationship, maybe I should think about dating somebody else." But when your wife or your husband does something you do not like, you say: "Oh well, it's her fault, or this bad thing happened but she's like that, forget it." That is the psychological immune system at work, helping you to see someone's behavior in a different way because you are stuck with it.

E.P. But if you are asked: What would you like? Which of these two states do you prefer? You prefer to be in the other one.

D.G. We usually say that, in general, no man would prefer to stay unmarried if he could avoid that.

E.P. This reminds me of bacteria. You know, in bacteria there are always a given share of them who are mutants; the ones who, sometimes, give up very good genes, just in case that future comes. Somehow you might find new genes that might be better adapted to the future. But if you ask bacteria—and nobody has asked them—but probably if we could ask bacteria whether they would like to be of the mutant variety, the ones changing, probably they would not. There is a fascinating question, well, a fascinating thing that

I always wondered and I am going to ask you straight. Some people say: "I want to go on vacation," and I say, "Well, I would go to Martinique." And then they go and look for a brochure, a leaflet on Martinique. They could ask somebody who has been on vacation in Martinique. But apparently they never ask anyone who has been on vacation in Martinique and who could inform them appropriately. They prefer to believe the brochure. Why?

D.G. Well, this phenomenon is something we have studied just in the last year or two. We have several experiments to demonstrate exactly what you have suggested. In ten years of research we demonstrated a number of errors that people make when they try to predict how happy they will be. The question naturally rises: "Is there anything that people can do to avoid these errors?" The answer is clear. Rather than mentally simulating your own future, why not find someone who has actually experienced the future you contemplate? We show in a series of experiments that people make much more accurate predictions about their own happiness if they base the predictions on experiences of others with similar experiences.

E.P. Instead of an advertising brochure!

D.G. Instead of trying to imagine what the future would be like for themselves. As you suggested, however, people do not want that information. We know from our studies the information will help them make more accurate predictions, but if you give them a choice they always prefer to imagine the future for themselves. Why? Because humans have an illusion of uniqueness. Each of us has. Because "I know the inside of my mental life. I hear your words, I see your face, I see what you do . . . but I know my thoughts and my feelings." And as a result, "I feel very unique in relation to other people. I feel different than the rest because I have such privileged private information about myself. Because I feel so different, so unique compared to others. I do not believe that their experiences can tell me all that much." I think this illusion of uniqueness is what stops people from using accurate information.

E.P. Is there any difference between men and women relative to this behavior? I think that—maybe I am totally wrong—most women think they are very different from the others. But do they feel more unique than men?

D.G. You raise a very interesting question. We never see stable gender differences, either in our studies or in the studies from other laboratories. We never see that men and women differ markedly in any of the behaviors or feelings we have studied. But one must understand that we do not look for such differences. If I were interested in gender differences I might try to dream up situations in which they were likely to occur and try to study them. So, the answer to your question is that we do not know gender differences, but that does not mean they do not exist.

E.P. The basic question now, the one which our audience awaits—I am sure. What about wealth? The search of wealth, of money, as a predictor of happiness?

D.G. Two very popular answers to the question "Does money buy happiness?" are yes and no. Our spiritual leaders tell us no, and those who are credited with having the answer, those who appear on television tell us yes. Both of these answers are wrong; they are too simple. Money does buy happiness, when it changes you from poverty to the middle class. Money does not buy happiness when it moves you from the middle class to the upper class. Let me examine these answers a bit. It is perverse to say that money does not buy happiness. If you go to Mexico City and see people who live in garbage dumps and do not realize that their lives would be improved with a little money, you suffer a great illusion. Your priest, or your rabbi, or your philosopher may try to convince you that poverty is unrelated to happiness. Money does make a big difference when it changes people's lives, when it provides safety, when it gives them food, and gives them shelter, when they do not have to worry about being hurt, when they do not have to worry about the weather, when they do not have to worry about medical care. But what we see—and this work is largely due to economists and not psychologists—is the relationship between money and happiness. At a certain stage of wealth, it just levels off. That stage of wealth in U.S. dollars, in 2004, is probably an annual income of fifty thousand dollars a year. So maybe an easy way to see this is that the first fifty thousand dollars you earn will buy a lot of happiness. The millions of dollars earned afterward do not seem to buy you happiness.

E.P. And maybe, if we compare this to the other basic factor, personal relationships, this money might explain why famous people have so many problems—apparently, as we read in the fashion news. Problems in their domestic lives and personal relationships, despite the millions of dollars they have.

D.G. A few curses of wealth are identifiable. One of them is that you—if you are like most people—you probably expect to enjoy increases in happiness when you make more than fifty thousand, right? "I made fifty million so I should be a hundred times as happy, a thousand times as happy," no? People feel very disappointed when they do not get more happiness from more money. Much of life is like that. You know, one glass of wine makes you feel good, two glasses of wine makes you feel great, but a hundred glasses of wine does not make you feel a hundred times better. No, you feel worse. So a curse of wealth is disappointment. It does not give us what we thought it would. Other curses of wealth of course exist. Studies of lottery winners, you know, people who suddenly come upon millions and millions of dollars generally are not happy people. A part of the problem is that they do not know how to spend the money. So money can make one happy up to a certain amount, but people tend not to spend it wisely. They tend not to spend it in the way psychologists suggest, in ways to give them the most happiness for their money. So you hit upon one result—we know that social relationships, consistently across the world, are the best predictor of human happiness. One of the best predictors of happiness is the quality of one's social relationships.

E.P. Better than wealth.

D.G. Much better than wealth. A very clear strong correlation means that you could maximize your happiness by the use of your wealth to give you more time for social relationships. Very few people—in my country at least—stop trying to earn more at fifty thousand dollars annual income and say, "Enough, I am just going to work half a day and spend the rest of the day with my family."

E.P. I am thinking in political terms, do you mean that, probably, what we tend to assume is a fine goal, that the politics of the left in wanting to raise the income levels of the poorest people is a better policy than, let us say, the right or center policies that aspire to raise all incomes—left, right, and center? Does what you find explain why one view is more popular than another?

D.G. I am not a political scientist—thank God. They live less happily than even parapsychologists. But certainly if you have control, if you are the single individual with complete control over the distribution of wealth and population, and your only goal were to maximize happiness in your population, you would prohibit people from earning more than the maximum amount of wealth they could possibly convert to happiness. So, of course, if there were a hundred thousand dollars in the world and two people, more happiness would be created by giving each of them fifty than by giving ninety to one and ten to the other. In a very few ethical senses you are quite correct. The politics that allow each individual to experience the early part of the curve, where happiness is being increased by money, surely is the best way to create the most happiness for the most people.

— 8 —

Psychopaths

Interview with Robert Hare

> In a utopian world, psychopaths would stand out as predators, because that is what they do, take advantage of others. We could live in a perfect utopia and there would always be psychopaths.
>
> —Robert Hare

In the history of evolution, in primordial times, more then one billion years went by without any predators because there were sufficient available resources to meet the vital needs of the first bacteria.* The time came when the nutrients ran out, even for that microscopic but proliferating population.

Robert Hare is Emeritus Professor of Psychology at the University of British Columbia.

*This statement is highly debatable. Predation by bacteria on other different bacteria evolved in the Archean era (see table 1) in organisms such as *Bdellovibrio*, *Daptobacter*, and *Vampirococcus*. Predation or not, there are never sufficient available resources for long in any population since all tend toward exponential growth. This was Charles Darwin's major point (Clark, 1981).—LM

The first solution was a more efficient use of the resources by means of symbiosis among different organisms specialized in specific functions; this led to eukaryotic cells. Billions of years later, some unwary inhabitants—or those without sufficient anxiety levels, as Kenneth Kendler would say—became the first prey.

With the appearance of the neural crest and then jaws in the skulled species, the predatory spirit was consolidated. The rest is very well known, except the reasons why violence continues in hominids even when it is not justified by survival, in the paradigm of natural selection. In this context, the most unusual and terrifying example is that of the psychopath, who symbolizes and unleashes the most incomprehensible processes of degradation and suffering. Robert Hare is the scientist who has best identified, and with greatest daring—because he dared to investigate, regardless of what was considered "politically correct"—the basis of the behavior of psychopaths. One renowned specialist said of him that "hundreds of thousands of psychopaths live, work, and play among us—your boss, your friend, or your sister—and it is possible that they will continue on their way toward destruction without being aware of it. And still more worrying: no one knows what to do about it, not even Robert Hare."

Eduardo Punset: Tell me about psychopaths.

Robert Hare: Psychopaths are great manipulators, and in the United States we describe them as being astute and ingenious, because they can deceive others, and even when you are an expert in the matter it is easy for them to deceive you and get their own way.

E.P. Have you been deceived?

R.Ha. Many times.

E.P. At the start of your career—

R.Ha. I began work for eight months as a psychologist in a jail. This was a way of earning money, while I was studying for my master's and Ph.D. I was the only psychologist in the prison, and I didn't know what I was doing. The first day at work I found myself with a psychopath and at that moment I wasn't aware of it. He entered my office and literally pinned me to the wall with his eyes. I didn't know what to do. To this very day they still haunt me as the most penetrating eyes I have ever seen. He took out an enormous knife and he told me that he was going to destroy things in the prison. I think he sought my reaction. I did nothing, I remained in my chair.

After my work there, I started a doctoral thesis on traditional theories of learning. I wanted to study how people could learn without the use of punishment. From my reading, I learned some psychopaths don't respond to punishment.

E.P. Thanks to your research, and that of others, we can now clearly distinguish the criminals and delinquents who are psychopaths from those who are not. Characteristics of psychopaths, you highlight, include lack of empathy, the inability to put themselves in the place of others, and the lack of conscience and of remorse. How is the lack of empathy or absense of remorse explained? Do you remember any examples?

R.Ha. There are many cases that demonstrated lack of empathy. It happens when you are not capable of putting yourself in someone else's place, more emotionally than intellectually. A psychopath can enter your brain and try to imagine what you're thinking, but he or she will never be able to understand how you feel. It's like trying to explain colors to someone color-blind. How can you explain empathy and emotions to a psychopath? A psychopath might be able to relate socially or intellectually in a significant way, but he or she conceives of people as objects. It's difficult to explain. Some people say that we all, as humans, think and feel the same way. That's not true. Police officers who impose the law are constantly in contact with violent criminals. At a conference I described the profile of a psychopath and the neurobiological bases of the disorder. The eyes of some police officers lit up because they immediately understood that some people they interrogate do not feel the same way they do. We believe that police, who are in permanent contact with murderers, liars, and the corrupt, are used to identifying psychopaths. It is not true. In the United States there was a very illustrative case. A man suspected of having murdered seven or eight women was going to be condemned for the murder of only three, because they were unable to prove the other crimes. To try to get him to confess, they told him to think of the suffering of the victims' families. Since he was a psychopath he was unmoved. He did not understand what they were talking about. In a lecture about psychopaths, homicide investigators were among the audience. Later, I learned from a judge that they had resolved the case because the culprit confessed. He confessed because they stopped appealing to his sense of right and wrong, his conscience and empathy, because he totally lacked them. They appealed instead to his sense of greatness. They claimed he was not a serial killer because he had killed only three people. If he had murdered another seven he would be a true assassin, they said. The suspect immediately confessed. The key is that the police realized that certain people don't think or feel the same as we do. This fact makes us all potential victims.

E.P. Psychopaths are also characterized by impulsiveness, incapability for future planning, irresponsibility, and irritability. They're always seeking new experiences that excite them. . . .

R.Ha. We have all seen people like that, though they may not be psychopaths. In addition to what you list, also notable is their profound lack of empathy and conscience. The characteristics you described work perfectly for everyone: there are husbands, wives, and parents who understand perfectly what we're saying.

E.P. Bob, you haven't yet mentioned another terrible thing that appears in your book. I did enjoy reading your books, I loved them, but I confess that they cause me certain anguish. Psychopathy, you say, is not innate. It does not develop in adolescence, but rather appears between the ages of three and five. This is still more distressing when you say that psychopathy has little to do with stable or unstable families. Your statistics, I think, show that criminals with a stable family enter jail at the age of twenty-five, whereas those from unstable families end up in jail at fifteen. But that is not true of psychopaths: it makes no difference whether your family is stable or unstable.

R.Ha. For psychopaths family stability is less important. For most people the environment in which they grow up is very important. They are forged as people based on the values of their family. But nature has given psychopaths something different from the rest of us. The normal forces of socialization that mold our personality, that make us more sociable and better citizens, do not work with psychopaths. The question is, why not? A psychopath can come from any family. If he or she comes from a violent family environment prone to social depravation, the psychopaths are accepted, because they are good apprentices. They rapidly learn the ways of committing antisocial and criminal acts without feeling the burden of conscience. Nonpsychopathic people brought up in this same kind of environment might end up as professional criminals and commit all kinds of dreadful crimes, but they have a certain conscience and don't feel good about their actions. People of this type are more loyal to other criminals and might have a good family life. Their problem is that they have learned to carry out criminal acts to get what they want. With psychopaths, on the other hand, it is as if they didn't have to learn how to be criminal, they simply are. Is it exclusively a genetic problem? No. Is it exclusively a problem of the environment? No. An interaction between genetics and the environment must exist, but the roles these two elements play and how they are distributed has yet to be explained. Psychopaths are people who, from very early on, have a certain propensity, not necessarily to develop in a certain way, but rather to personal characteristics such as the lack of fear or anxiety, a liking for an easy life, and the tendency to be impulsive. They are individuals who cannot be inhibited or molded by their social environment in the same way as normal people.

E.P. When we see young people being cruel to animals, who easily lie, refuse to obey their parents and teachers, and do not fear punishment, what can we do?

R.Ha. We must be careful. The vast majority of these children will grow up to be completely normal. We don't know what the specific combination of dangerous characteristics is, though we have identified many of them. What can be done? First, we have to understand that the illness does not suddenly appear at the age of eighteen. What happens before that? Suddenly they are eighteen years old and they are psychopaths—it doesn't make sense. If we are not capable of recognizing it, we are condemned to treat exclusively adult psychopaths.

E.P. You produced a manual to identify psychopaths that is used around the world, the PCLR ("Psychopathy CheckList Revised").

R.Ha. The PCLR is a tool that was created in order to diagnose psychopathy as objectively as possible. At first, it was only a research instrument. I never imagined that it would be used in criminal justice, or to determine who would be let out of jail early and who would receive treatment. It was developed exclusively as a research tool to allow other investigators to have a common measurement that we could all use.

E.P. If the PCLR is applied to everyone, we arrive at the figure of 1 percent of the population that suffers from psychopathy, the same rate as schizophrenia, right? In the United States alone, for example, this means two million psychopaths.

R.Ha. It's more than we thought, especially if we consider its impact. Just think of the number of people affected by schizophrenia—the family, close friends, and the person him- or herself—who suffer a great deal. Psychopaths feel no personal anguish nor do they have any problems. You have the problem. They affect hundreds or even thousands of people so their impact on society over the course of their lives is much greater than simply the number of cases would indicate.

E.P. It's incredible. What is the basis of psychopathy from analysis of the brain? Psychopaths have strange brains, different from those of other criminals, but in what way?

R.Ha. The structure of their brains seems to be the same as normal people. Their functioning is different. We have been able to analyze the parts of the brain that are active when people perform certain tasks or process certain information. A most important finding in recent years is that when a psychopath tries to analyze something with an emotional charge, such as photos or words, the parts of his or her brain that are activated are not the same as those activated in the majority of people.

E.P. You mean that if you mention a negative emotional response, for example, rape, to a psychopath, the area of his brain that processes this is different from that activated in normal people?

R.Ha. Exactly. If we show the word "rape" to a psychopath, he or she understands it as a neutral word, like "table," "chair," or "tree." There seems to be very little difference in the way they respond, or in the parts of the brain that are activated. We have carried out several experiments in which we show very unpleasant images to psychopaths: crime scenes. The functioning of their brains shows that they respond as if the scene was something normal, a dog or a tree. Parts of their brains are not activated. They are the limbic regions of the brain, the parts of the brain associated with the processing of emotions. In other experiments we discovered that certain parts of their brains related to language are activated. Another characteristic of psychopaths resembles the characters of Doctor Spock and Data in *Star Trek*. They say that something that arouses emotions is "interesting." Psychopaths analyze things linguistically and not emotionally.

E.P. You also suggest that instead of the use of the left hemisphere for language, like most of humans, psychopaths use both hemispheres at the same time, which generates a kind of competition or frustration. But why does this lead to psychopathic activity?

R.Ha. We discovered, in 1994, by use of a laboratory device, that psychopaths process language in both hemispheres. This is not very efficient, because one of the sides has to be in charge of everything. It was discovered that psychopaths process emotions in both hemispheres, instead of processing negative emotions in the inner right hemisphere in the front part of the brain.

E.P. Let's talk about therapy. You suggest that rehabilitation programs for psychopaths won't work—that their brains function in such a way that they simply learn how to deceive. You say it is best not to try rehabilitation using conventional programs. How can we say that to the judges, people in charge of prisons, teachers, victims?

R.Ha. Everything has been tried but nothing works. I know this is dreadful but we are not about to give up. You go to the doctor with a headache and he prescribes aspirin. Then you return with a stomachache or a broken leg or because you suspect you have something serious, but whatever it is, the doctor always prescribes aspirin. You must change your doctor or you will die. This is true of the criminal justice system. Why should a treatment program work with all delinquents? As you said, traditional rehabilitation programs are very little use with psychopaths. Two studies demonstrate that delinquents that follow these programs commit more serious crimes than if they had not been treated. The programs don't make the situation worse. The programs were not the right ones and the psychopaths learned new ways of manipulating people.

E.P. Genetics and pharmacology have established that aspirin does not act in the same way in all people. While it works very well for some people, it kills others; it depends on genetic variation.

R.Ha. Precisely. The interaction between the treatment and the nature of the individual is crucial. In most treatment programs this fact is not taken into account. We tend to treat everyone the same.

E.P. Bob, if for now nothing can be done to correct the lack of conscience and empathy, and therefore no one can be changed, as psychopaths lack these two fundamental qualities, and you say that at the age of forty they become less violent and psychotic . . .

R.Ha. Less psychotic, no. Less violent, less impulsive, and with less need to call attention to themselves.

E.P. Can anything be done?

R.Ha. Your question is absolutely correct. Spot on. Those two basic aspects of our makeup, lack of empathy and conscience, don't change. They are personal characteristics that don't vary in the course of our lives. We have a lot of data that demonstrates that lack of empathy, their delusions of grandeur, and their emotional shielding are characteristics that remain throughout their lives. To a certain extent we can change their impulsiveness, their need for stimuli, and perhaps their irresponsibility. The change is not spectacular. Years ago, Lee Robins wrote a famous book entitled *Deviant Children Grown Up*, which explains that they may improve with age, but they continue to be pretty unpleasant. But I don't want you to think that there's nothing that can be done. In several countries, such as New Zealand, the United Kingdom, and Canada, they work on programs designed specifically for psychopaths. I work with a colleague on a program that might work. It doesn't appeal to their sense of conscience or empathy. Rather it is based on cognitive behavior. We must try it. Otherwise all we know how to do with psychopaths is lock them up. Many people like this: Since we can't do anything, let's lock them up. But this presents a danger for everyone. We want to develop appropriate programs, which will probably not have a drastic effect, but at least they will reduce the propensity to violence.

E.P. My last question is almost irrelevant after our conversation. Why have experts taken so long to differentiate the behavior of psychopaths from that of other criminals?

R.Ha. The question is not at all irrelevant, it's a very important question. I think we tend to think that other people think like us. We do. We think people are inherently good. We believe that if you give them an opportunity, everything will be fine. Many people in the United States think that if you give them a puppy, a hug, and a musical instrument they will improve. It's not true. It's not that easy.

E.P. It's difficult to accept that not everybody is inherently good.

R.Ha. Not that they're inherently bad. Some are more difficult to mold, socialize, than others. And psychopaths are among the most difficult to mold. But one of the greatest complications in treating this disorder in particular is its difficulty in recognition. I gave a presentation in Wales called "Snakes in Suits" about the profile of the psychopath who is not in jail but rather forms part of the management team, as a sales representative, a husband, or a wife. People we don't identify as psychopaths, although their victims do. After such a presentation people comment, "You have described a person I know who is not in jail." Governments and society want an easy answer, as if the problems were only financial and social. It's not enough to invest a lot of money. In a utopian world, psychopaths would stand out, as they would be predators, because that is what they do, they take advantage of others. Even if we could achieve a perfect social utopia, the psychopaths would not disappear.

Anxiety

INTRODUCTION

All life on Earth at all times and under all circumstances requires an appropriate flow of matter and energy to grow and maintain bodies. All life requires a source of expendable energy. Living beings need the continuous flow: we all need matter in the form of the chemical elements, carbon, hydrogen, nitrogen, oxygen, phosphorus, and sulfur. But just these elements (and the requisite few salts such as that of sodium, potassium, and chloride) are not enough. The form, that is, the chemical composition of compounds made of these elements and the presence of the compounds in liquid water, is absolutely essential to the continuity of all life forms. The absence of water, recognized as thirst, and the absence of adequate carbon compounds, detectable as hunger, are unmistakable and immediate signals of the inadequacy of appropriate matter and energy flow at any given time. Interruption in energy and matter flow, or even inappropriate timing of the flow, are universal phenomena that predictably lead to stress. So does the perceived presence of a predator, a nonliving environmental threat, even our own shadow. We humans are especially adept at the interpretation of such signals, signs, or symbols that the necessary energy and food are absent, delayed, or unavailable. We seek signals, signs, symbols of reassurance that our energy, food, and water are on their way. We are "interpretative mammals": if flow of these necessities is halted, restricted, or we perceive credible threats, we suffer stress. Stress is inevitable and all of us, of course, recognize its familiar signs. Although we human animals are especially adept at the interpretation of signals, signs, and symbols of imminent interruption in our energy, food, and water flow, we often, upon study, realize that in retrospect, our detailed misinterpretations may be clearly recognizable. Our misinterpretations may lead to further stress, to paranoia, to catatonic states of being, to anger, to music, or at least lyrics. Only recently has stress been a subject of direct study by university faculty members. As usual, the scientific studies are always limited, based on assumption. They seldom warrant generalization without verification. It is especially pernicious when laboratory and field investigations are overgeneralized to generate the hype and exaggeration so prevalent today in the public media. Our interviewees are cautious. It is probably safer to believe what each says here than any TV or newspaper report. But only after—please—reading the references to the scientific findings to which they refer. In any case, be aware that humans represent only a minute part of the biological universe of entities capable of being stressed. So please read with skepticism as Eduardo Punset asks these scientists and scholars about anxiety and consciousness studies that interest him.

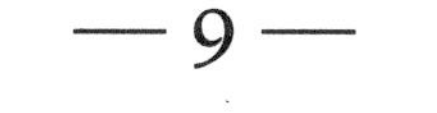

— 9 —

No One Is Boss

Interview with Daniel Dennett

Cells do not know who you are, nor do they care.
—Daniel Dennett

Chimpanzees and humans are the only animals able to recognize themselves in a mirror. They may be the only ones to possess consciousness of their own. Dolphins, not fish but mammals, may also. But where and when in the history of evolution does personal consciousness appear? When we try to describe consciousness, do we illustrate the soul? We seem closer than ever to discerning at least self-consciousness, thanks to the impressive team of scientists that includes Richard Dawkins, Francis Crick, James Watson, Edward O. Wilson, and Steven Pinker, directed by Daniel Dennett from the Center for Cognitive Studies at Tufts University near Boston.

Daniel Dennett is director for the Center of Cognitive Studies at Tufts University.

We had better understand the fierce debate of those who question the dual nature of humans. Both factions seem to have irrefutable views, formulated with an elegance and clarity rarely found in scientific prose. It is a profoundly ambiguous debate. The most reductionist scientists, like Dennett, reserve space for consciousness, while the staunchest defenders of the power of the mind, such as Deepak Chopra, see science, not religion, as the preferred mode to understand nature.

The commotion stirred by the denial of the existence of the soul by the scientific community is not surprising. The outcry raised by antireductionists was generated by their profound fear of existence without free will. French physicist and mathematician Pierre-Simon Laplace summarized with clarity the essence of determinism. As flies are dazzled by light, humans debate, unaware of determinism at times. So often discomfort leads to loud rejection of the concept of a universe undefined.

Eduardo Punset: So this machine, which is our body, has no spirit?

Daniel Dennett: Exactly, it does not have any spirit.

E.P. Nor soul.

Da.D. There may be a soul. As far as a spirit is concerned, let us take stock of the human body, and what do we find? Several billions of different cells, living cells, neurons—cells of every type, in other words. But none of the cells know who you are, nor do they care. But in some way, you join the team of trillions of cells, rather fascist cells, which, like slaves, expel intruders as bees do in a hive, and they do not care for democracy, Barcelona, or Boston. Our cells are unfamiliar with all of this, and yet, here are two great teams of cells, your team and mine, that know a lot of things of which they are not aware. Therefore, it is natural to think that the only explanation is that, in addition to all this tissue, there is a spirit in the body, and that a soul does exist.

E.P. It is what people believe.

Da.D. It is a very powerful idea, but it is wrong. A spirit cannot exist in a machine.

E.P. Dan, if there is no soul in the machine body, then what governs it?

Da.D. Let me illustrate it using a fable. In a country there reigned a king, and without the king the nation did not exist. But then democracy was invented. Something similar happened in biology, even though biology invented democracy first. No part of you is the king, there are only factions, political parties, and groups that compete for control of your brain.

But what sort of groups are they? They are groups of activities, they are not cells but rather information models that compete for control of your body, and because competition is dynamic and fluid, there is always someone at the controls, someone in the office.

E.P. Even though they have not been officially elected, right?

Da.D. Exactly, there are no official elections, but the brain has an opposition process that allows one issue after another to surface, to go forward and take over the situation. While we talk, we are a collection of ideas, projects, hopes, and plans that we might put aside at the end of the day, and a different me will take over the controls. All the projects are linked by memory. There is no one spirit, but rather a succession of temporary controllers.

E.P. What does consciousness mean? What I mean is, how did it all start? How can we explain it? How can we know what exists in this machine?

Da.D. To my knowledge, the key to understanding consciousness is to ask the evolutionist's question: why? Consciousness exists so that we may anticipate the future and have foresight in order to plan and project a better course in life. In a human that has a background of language, science, and culture, there is a great deal of information being exchanged. We can predict eclipses that are not expected for another few centuries, imagine distant galaxies, and think of a past we never lived. Consciousness is the power of the brain to represent things that do not occur in the present, strictly speaking, but rather in the past and future. Our consciousness is that power, it is the physical world's capacity to represent itself.

E.P. And what genes will carry the information from one brain to another?

Da.D. We call them memes.

E.P. Memes?

Da.D. Yes, I use Richard Dawkins's term, who says that memes are like genes; they are the units that transmit and replicate culture.

E.P. Like traditional genes, right? Are you suggesting that for memes, or cultural genes, to exist, there has to be a capacity for communication among different brains that did not exist before?

Da.D. True, language is very important, but it is not the only way. There is also a type of cultural transmission in other species. For example, consider the methods used to build nests or dig burrows. It has been discovered that chimpanzees have a way of catching termites with a stick, which involves using a sponge to take out water from the trunk to drink. It is a minor form of cultural transmission compared with that of humans, but it is still there.

E.P. Do you mean that it is not simply a different grade, but that it is completely different?

Da.D. In biology, you always start with a different grade of complexity, which increases from there. Therefore, for us culture is not a question of thousands but of millions, and not only grades. It is a million times greater than that of the chimpanzees. For example, were we to grow up in a desert island without the influence of human culture, we would not be that different from a chimpanzee. Who would invent arithmetic, language, cooking, or agriculture? These skills, abilities that are so distinctly human, have been acquired through cultural transmission, and that is what composes our mind. If you think about human beings, you realize that their minds are not just in their brain, but also in the library, in computers, in friends, and in all the tools used to think, acquired throughout their lives. If you take away these tools from a person, you make them quite vulnerable.

E.P. Going back to my own obsessions, Dan. Who looks after the brain, assuming that someone looks after it?

Da.D. It is an idea that is a little scary. The brain has ten billion or perhaps one hundred billion neurons, and that is not all. Not one neuron is aware of who you are, nor does it care. They are too stupid for that. Therefore, it requires a democracy; neurons work in teams and they compete among themselves without being overseen by anyone, because nobody can oversee them. The theory that the consciousness or conscience (as you know consciousness is the same word as conscience in Spanish) has a supreme master is simply wrong.

E.P. It would be like walking into a factory devoid of people.

Da.D. Exactly, like an automated factory; full of machinery, but empty of personnel. There are parts of the brain that do act as agents responsible for following up other parts, and therefore are like civil servants, but without consciousness.

E.P. They cannot think—

Da.D. No, it is you that generate thought, not the parts from which you are made.

E.P. Therefore, if someone wanted to investigate powers of the brain which are unknown, invisible, and intangible, we should have to tell them they are wasting their time.

Da.D. We do not even understand the most elemental abilities of the brain, in fact, such as its ability for multiplication, walking, or telling the difference between a cat and a cup. These questions may be difficult to answer, but I do not believe the idea that the mind has mysterious powers helps. I believe, however, that with time we will be able to explain the functions of the brain in normal biological terms.

E.P. If we observe the universe and we try to gain knowledge, where do our senses come from? Does anything exist beyond our own physical perceptions? How do we attribute meaning to anything? No matter how much we progress, we still have not attained what Newton said: "I would like to know the mechanisms by which a visual perception of the universe is transformed into the glory of colors." Remember? Dan, we do not understand something as basic as that yet; it reminds me of the vain search for the precursor molecule of the first cell.

Da.D. In the first place, bear in mind that we are the only species that poses those sorts of questions. Cats and dogs do not sit around meditating on the meaning of life, nor do they ponder their position in the world. Their states have meaning because they have been given a sense for the history of evolution from where they emerged, and their own experience. A dog knows that a certain visual situation means danger, such as if a man once hit it with a stick and the dog then recognizes the man. This is where its sense comes from. The dog's brain is designed to be able to extract this sense from memory and apply it to certain situations. Our brains function in the same way, but because we have the ability of language, we can reflect over events and construct layers of meaning upon the sense derived from our biological makeup.

E.P. Are you then saying that there is no such thing as extraordinary perception, no extrasensory perception, nothing beyond the neurons and atoms?

Da.D. Yes, it does exist, but it does a marvelous job of hiding, because there is no actual evidence.

E.P. Therefore the brain has no prodigious capacity.

Da.D. The capacities of the brain are prodigious. A child that learns a foreign language, or goes to school, is showing the prodigious capacity of the brain, which has not yet been fully understood.

E.P. Then what do you think of people who believe they have identified supposedly occult powers such as thought transmission or time travel? I know it must seem preposterous to you—

Da.D. People who claim such things probably believe them, but it is not true. Pablo Picasso once said, "I do not search, I find." This "I find" is a great explanation of what is an alert genius, even though I do not think it is true. Picasso was indeed a genius. His discoveries, however, were not miracles, but the fruit of much research previously carried out by him, and going back to our sphere, natural selection.

E.P. What role does intuition play then? Einstein said that, as a last resort, discoveries were born from a great intuition.

Da.D. That is true, but what is intuition? When we get a brilliant idea and we say we came to it through intuition, in reality we acknowledge we do not know how we came to get that idea. In other words, we say it occurred in the brain, through the workings of our neurons, competing among each other, suggesting ideas. Some people's brains are better than others' in generating new ideas, probably as a result of their acquired thinking habits. In fact, they teach themselves to think differently, but there is nothing extraordinary in that.

E.P. Therefore, are you saying it is an exercise of the will?

Da.D. It is like a magic trick. If we see a magician perform a magic trick on stage, at first we think it is extraordinary, impossible, and magic. Later, when we understand how it was done, we say it was nothing more than trickery. Science is proving that something similar occurs in the brain, which can perform dramatic tricks.

E.P. Including intuition—

Da.D. Definitely, including intuition.

E.P. What can be said about consciousness? You are a highly renowned expert, and you have reflected long on the idea of consciousness, even though they accuse you of being a reductionist. Have there been any new discoveries in this field since we last saw each other a couple of years ago?

Da.D. I think we have made great progress where consciousness is concerned. We are dividing it into its constitutional parts. With that, we are gradually unearthing the idea that consciousness is a grand and mysterious thing, while we realize that it consists of all these parts and that there are grades of consciousness. If we examine bacteria, we would say that they have no consciousness whatsoever, that one bacterium is merely a small robot. Brain cells are identical of course, because they are also small robots. If we observe a tree, we can believe that it is not conscious. But it is alive; it may not have a consciousness, but it is sensitive: it distinguishes the world around it and measures whether it is too cold or hot, if it has enough water or wind.

E.P. Or sunlight—

Da.D. Yes. It feels the lack of sunshine and turns toward the sun. It is a type of consciousness, a certain sensibility. Many people believe that you cannot use this sensibility as a basis to explain the sensibility of a tree to sunlight, or the response of the rod and cone cells of our retina to light or to build a theory of consciousness. Why not? Because when all is said and done, the lens or film of a camera are also sensitive.

E.P. But neither lenses nor films have consciousness. . . .

Da.D. But our retina itself has no consciousness either, and yet on the basis of its sensibility, it reacts when faced with light-dark differences. This consciousness is constructed.

E.P. Some biologists ask themselves whether bacteria possess consciousness. It is amazing to see bacteria move as they follow magnetic currents, as though they were conscious.

Da.D. In fact, the speed of movement is very important. When David Attenborough produced this marvelous television series on the secret life of plants, called *The Private Life of Plants,* he shot wonderful sequences that showed how the growth of plants accelerated during the filming. It showed how the vines grew, twisted, and strangled trees—

E.P. Consciously?

Da.D. It looked as though the plants were conscious. But the effect is produced simply as a result of the time of the reaction. If we were to slow down a human to the rate of the growth of a tree, he would be very drawn out. You would think that this person was not in the least conscious, like a robot. We say that robots are not conscious, because robots look stupid, saying things like "I-AM-A-RO-BOT." But robots like C3PO and R2D2 from *Star Wars* seemed conscious, because they moved correctly at the same pace we do.

E.P. Speaking of velocity, light travels faster than sound, but apparently the brain processes visual images much more slowly than auditory perception, so at a distance of ten meters, for example, sound and visual image reach the brain at the same time. I would like to ask you two questions: First, is that simultaneousness absolutely necessary for the brain to interpret what is happening? And secondly, seeing as there is no boss in the factory which is the body, is there a point where this light, this entrance of visual image, and sound meet?

Da.D. There is more than one point where signals meet. There are many, and it happens at different times. We need to understand that there is no place in the brain where the arrival of signal matters. There is no place where the stimuli act to concur "now that we are all here, we have crossed the finish line and here we are at the place of consciousness. There might be two points, ten, or two thousand. And in one part of the brain, light information arrives before sound information stimuli for the same event. In another part the stimuli arrive in a different order. The brain may be able to organize, to reorder, to process all the stimuli; but it can be tricked. No official office space represents "Here is consciousness; the buck stops here." Therefore, the question about what process took place first in the conscious perception, light or sound, is a poorly asked question that has no answer.

E.P. What happens when the brain tells your toe to wiggle while it tells your eyes to look at a painting in the room? The distance between the brain and the toe is greater than the distance between the brain and the eyes, therefore the timing could be off. How does the brain manage to make them coincide?

Da.D. The brain has had a lot of time, all the time of evolution, to find a solution to this "stimulus" problem. And it found it. When the brain initiates behavior by sending motor signals it also sets time expectations. Certain results are expected after some given time. If you send a letter from, let us say, Boston, to someone in California, you do not expect a reply the next day, but you do by the end of the week. The brain can be tricked by signals that arrive too quickly, and thus create an abnormal experience for the brain, because of its own internal expectations of time. In the end, the only way to confuse the brain with regard to simultaneity or lack thereof is by contradicting and canceling its own expectations. As long as events do happen as the brain expects them to happen, everything is fine.

— 10 —

The Hidden Self

Interview with Oliver Sacks

> I assumed that the memories I did have, especially those which were vivid, concrete, and circumstantial, were essentially valid and reliable. And it was a shock to me, but I found that some of them were not.
>
> —Oliver Sacks

Eduardo Punset: Your library is full of Oaxaca: cuisine, plants. What did you love in Oaxaca?

Oliver Sacks: Although I went to Oaxaca especially to look at flowering plants

---

Oliver Sacks is clinical Professor of Neurology at the Albert Einstein College of Medicine, Adjunct Professor of Neurology at the New York University School of Medicine, and Consultant Neurologist to the Little Sisters of the Poor.

and ferns with my fellow botanists, I was interested in everything. I love the food there.

E.P. Ferns. We call them *helechos* in Spanish. Does your passion for *helechos* have anything to do with those plants, which are millions of years old?

O.S. They are great survivors. Dinosaurs came and went, but ferns are still going. It also goes back to my own life; before the war our garden was full of ferns. My mother loved ferns.

E.P. Ferns are not especially beautiful.

O.S. They do not have the multicolors and exotic forms of flowers. They are a much simpler form of life with their own beauty, a very delicate beauty.

E.P. They are survival machines, and you have really looked as nobody else into different patterns of survival from neurological disease. After so many years of this passion, do you really think we are learning anything about this age-old brain? Are you and your friends, Antonio Damasio and Joseph Ledoux, really discovering something new?

O.S. The human being, the organism with the big brain, is tough and versatile. When a machine is damaged, the trouble may be irreversible. When a part of the brain is damaged, often another portion of the brain can take over. There is great reserve and plasticity in brains. If you lose any one way of doing things, the brain—you, in other words creates—or discovers another way to do the same, or nearly the same, thing. I also think the person, the identity, is very tough biologically and tends to survive all sorts of damage.

E.P. For the sake of our readership let us remind them a little bit about what is probably your most famous contribution, the description of "The Man Who Mistook His Wife for a Hat." This was a case of visual amnesia, or whatever you call it. What happens in the brain of someone who mistakes his wife for a hat?

O.S. This man, Dr. P., was a great musician and singer. Because the visual parts of his brain began to degenerate, he had difficulties in recognizing people and places by sight. As soon as he listened to people or touched them, he would recognize them. His hearing and sense of touch remained intact, but he started to make absurd visual mistakes. On one occasion, when he reached out his hand for his hat, he got his wife's head instead. For a moment he made this outrageous mistake, which eventually became the title to my book.

E.P. Through patience do you think something can be done about this sort of error? Can a patient like this be helped?

O.S. Dr. P. could not find his clothes, and he was unable to do all sorts of things. But he eventually was able to replace deficiencies in visual recognition by musical recognition. If he sang and if he converted visual clues into a piece of music, he could briefly recover his normal orientation. I have been seeing another patient, also a gifted musician, with the same visual problem, although she still has a very acute sense of color. In her apartment, everything is organized by color, shape, position, and association, so that she can find her way around the world, because she has reclassified it in abstract visual terms.

E.P. She has reconstructed the world to her convenience.

O.S. These examples show the adaptive mechanisms of the brain.

E.P. It seems that in the most widespread brain damage, the perception of music is the last thing lost. Recent studies of the prehistory of the brain show us that it is uncertain which came first, music or language. Language is located in a particular part of the brain; however, it seems that music is spread throughout the brain. Why do you think music is so important in the brain and for the human being?

O.S. A very crucial question. My colleague Steven Pinker speaks of music as auditory cheesecake; he means that it is irrelevant. Certainly its representation in the brain is very widespread, so that pitch, rhythm, and other aspects of music are all organized in different parts of the brain. Even when there is very widespread damage people always respond to music. I think that music, language, and mimicry are the three exclusively human attributes, which are species specific, and perhaps they evolved together. I do not think our music is comparable to animal calls, bird song, or whale song; yet I think it did develop in a social setting, in an emotional setting. It is present and important in every culture and for every individual. I would go mad if I were deprived of my piano, or could not play music.

Music has great organizing power. A sequence, which cannot be remembered as such, can be remembered if it is organized into music. Often people with aphasia—who have lost language—will be able to remember language if it is imbedded in music. People with musical hallucinations suddenly can "hear" music with such vividness that they think a radio is playing or that someone is playing the piano in the next room. This is different from imagining music, as they think they are perceiving it. It is a surprisingly common disorder. It is especially common in people who suffer hearing loss, but it occurs in other conditions, as well. I think the human brain is very, very tuned to music, even in so-called nonmusical people.

E.P. Your first therapeutic experience with music was with that marvelous book, film, and story called *Awakenings*. That is where you perceived the possibilities of the use of music to reconstruct the patient's world.

O.S. I would see patients who were motionless or frozen, and some perhaps who could not even walk or take a single step. Yet they could dance. They would be unable to utter a syllable, but could sing. Music would somehow seem to bypass, at least for a few minutes, their parkinsonism and to liberate them into free movement. Everything would change; their brain waves changed. There is a profound neurological change with music.

E.P. People have wondered a lot about that almost fanatical "boy chemist" who was always thinking about chemistry, and who then developed into this physician, obsessed by neurology and your patients. You have always been reluctant to talk about this change from chemistry to neurology, from objects to hearts, as you were saying. Have you reflected on that?

O.S. I do reflect on it sometimes, and it is not just reluctance, it is actually bewilderment. I do not actually know what happened. However, if I explore it more, I might write a companion book, a second volume to follow my *Uncle Tungsten* (Sacks, 2001). But certainly for me the first world was one of objects, although a very important early world was the world of plants. Plants at least are alive, and they are organisms. You have to have a feeling for plants, although plants probably have no feelings for you. Then later I became interested in zoology and in nonhuman animals. Finally human beings interested me. Those early interests in matter (chemicals, physics, and botany) are still there underneath. This sort of slow movement is like the movement of evolution itself. The Earth starts as an inorganic body, then its animals and plants, and finally people, evolve. Sometimes I wonder if my choice of chemistry was an early trajectory, which did not get anywhere. Now and then perhaps one has to make a new trajectory. Six hundred million years ago there were some very strange organisms called the Ediacara biota—some look like animals—but they did not survive. They left hardly any traces and all died out.

E.P. But they were beautiful.

O.S. Yes, they were very beautiful. Some paleontologists say that Ediacara was a failed experiment of evolution. Maybe chemistry was my Ediacara, maybe it was the essential prerequisite. But chemistry, even the history of chemistry and its personality, has always fascinated me. For example, tungsten was a favorite element of mine. It was important for me to know that metallic tungsten was discovered in Spain in the 1780s by two Basque brothers, Juan-José and Fausto Elhuyar. I have always wanted to know the human aspect of science and technology as well as the science itself.

E.P. Speaking of tungsten, I see that you have two weights there.

O.S. I collect paperweights; I have them made of almost all the chemical elements. This is a regular paperweight, and this is tungsten, or wolfram. I love its density. Maybe that cylinder will still be around in a million years. It may be a survivor, and whether people, or plants, or chemical elements, I like survivors.

E.P. A great American physicist told me that when you see a picture with a tree in the foreground and some rocks behind, people tend to think that the tree will disappear and that only the rocks will remain. But even rock, even tungsten may disappear. It is life that will not disappear. This is a nice idea, don't you think?

O.S. Life is very tough. Life probably began as long as four billion years ago, as soon as there was liquid water on the Earth and the temperature was cool enough. We share the DNA of the first living organisms. You are right, rocks come and go, but life has a different sort of toughness. Of course, the weather or physical circumstances erode rocks, they crumble and dissolve, but life has continuity, the genetic message is passed to the next generation. Life transmits its genes, its DNA. We are connected by all our ancestors, back to the first ancestor four billion years ago. Tungsten does not have ancestors the same way life has, although it is difficult to destroy an atom. The first identity, in a sense, is the identity of atoms, so tungsten is always tungsten, always a particular atom. My interest has always been in identity. Of course genetic identity and then personal identity has interested me. So maybe, for me, my whole intellectual and emotional trajectory has been about identity.

E.P. True, true. You have maintained interest in the chemical elements, the basic elements, since childhood. You have periodic table examples from the pieces all around your office, in your memory, in the museums where you used to go.

O.S. I thought you were going to mention the periodic table hanging in the bathroom. Did you not see it?

E.P. Yes, I did, and somebody from my film crew said it was the first time in his life that he had seen a copy of the periodic table of elements hanging in the bathroom!

O.S. If he went into my bedroom, he would find a huge periodic table on my bed. I call it "sleeping under the elements." My quilt is decorated with a complete copy of the periodic table of elements.

E.P. Has this reassured you that there is some sort of order in the chaos in which the universe expands?

O.S. Yes, I think very much so. When I was a little boy during the war, like most of my generation, I was sent away from London, from my family. The school I was sent to was chaotic and cruel, nothing and nobody could be trusted, and nothing seemed predictable. When I came back to London, I had this uncle whom we called Uncle Tungsten because he made filaments for bulbs out of tungsten. He introduced me to tungsten, to chemistry, and to the periodic table. From him I acquired a sense of order in the universe. Perhaps the relative predictability, the dependability of elements and their behavior and their chemistry, all this had to come first before—as you say—I could be reassured. I still love to touch tungsten; its solidity gives me great pleasure.

E.P. It is marvelous. You and other neurologists have made a major contribution, a revisionist contribution in that you never accepted the traditional idea that memory is just a sort of recording of daily experiences in the cortex. What do you think exactly?

O.S. I probably did accept the classical notion that compares memory to a sort of recording or photography. About twenty years ago, I started to think of memory in different ways, as something one constructs; one assembles from different elements, elements that are never the same. I have been writing a little about this now in relation to my autobiography, *Uncle Tungsten*. In that book, I describe two vivid memories of bombs exploding in London when I was a little boy of six. One of the memories I described to an older brother after the book was published. He agreed, and said it was exactly how he remembered it. With the other memory of bombs in our back garden, however, he said to me, "You never saw that." I replied, "What do you mean I never saw it?" He said, "We were away at the time." I said, "But I can see it now!" I could see the bomb exploding, and saw my brothers bringing pails of water and the bomb shooting out hot metal. "How come I can see it now?" I asked. And this brother replied that our older brother had written us a letter, a very vivid letter, and that I was fascinated by his description. Unconsciously, I constructed the scene from his description in my mind. Then I appropriated it. Now I know this intellectually, but even so, I cannot distinguish the true memory from the false one. One seems like the other, and I think something like this shows both the strength and the weakness of human memory and human imagination. One makes up things often without being clear as to what is their source.

E.P. In a recent article, and I quote you, "I assumed that the memories I did have, especially those which were vivid, concrete, and circumstantial, were essentially valid and reliable. And it was a shock to me, but I found that some of them were not." Is this related to the idea we have of the evolution of the brain?

O.S. We now think in terms of networks in the brain. If one part of the network is knocked out, another part can take over. But the networks in the brain spread into the cultural network, and so much of our memory is externalized. All the books in my office, all the recordings, everything I hear on the radio, one's whole culture, is external memory, which one incorporates. This gives great strength since it is not just our own experience that creates us, but everything in our culture. At the same time, we are not always sure of the source of any memory. A good colleague of mine talks about the fragile power of memory. No two people will describe an incident in the same way. Witnesses of a crime will give different stories. None of them are lying; they make their own associations, and they have their own emotions. In the 1890s, Freud was puzzled when so many of his patients would give him descriptions of having been sexually abused in childhood. At first he took this all to be literal historical truth, and then he started to wonder sometimes if imagination or fantasy might have entered into these narratives. For example, people describe being abducted by aliens, or of being taken up into a spaceship.

E.P. How can a patient come to the conclusion that his leg is not his leg, and that someone is just pushing that thing around?

O.S. This is a very easy conclusion to come to. I wrote about such an experience with myself, and well, it astonished me. I had an injury in 1974. I fell on a mountain hike and I severely damaged the muscles and nerves of my leg. I had to have surgery, and the leg was immobilized in a cast. But then a strange episode occurred when, two days after surgery, the nurse looked through my door; she rushed into my room and said, "Dr. Sacks, watch it or you'll have the leg off the bed." And I said, "What do you mean? It is in front of me!" Then she said, "No, your leg is half off the bed." I said, "Look, shouldn't I know where my own leg is?" She said, "Yes, you should." I said, "Well, I do." She said "Look." And I looked. My leg in its plaster cast had fallen half out of the bed while I was asleep.

E.P. And you did not know this?

O.S. No, I did not. This drew attention to the fact that I was no longer receiving any sensation or any information from my leg, and therefore I could no longer be certain where it was or indeed that it was. I could only recall a previous visual memory. We depend on continuous information that comes from our bodies to the brain. If that information is cut off in any way . . . one of the things I say in my book, *A Leg to Stand On*, half as a joke, is that the reader will understand what I mean and will have a similar experience if he has a spinal anesthetic. If he undergoes a spinal anesthetic, he suddenly feels that he terminates at the torso. What lies below the torso—a pair of legs—is not his. They look like some sort of strange anatomical models, which, although attached to him, are not his. This situation, to be believed, must be experienced.

E.P. This is terrible what you just said—that your own body only really exists for you if there is communication between the brain and that part of the body. Otherwise, it is like a nonentity, is not it?

O.S. A former student of mine, Jonathon Cole, wrote a beautiful book about this. There are some conditions, sometimes infections, which completely take away all sensation of the body's location in space. At first this can be completely disabling: you cannot sit and your limbs are everywhere. It is almost worse than being paralyzed. But we talk about survival. This nineteen-year-old patient of Dr. Cole's lost all sensation from his face down. Yet now he is able to stand and drive a car and do all sorts of things even though he has had no neurological recovery. He does these things using his eyes and by planning. His is a wonderful example of adaptation.

E.P. One facet of your work fascinates me. I am aware that you started writing from a very early age, and apparently, like Picasso with his painting, you never stopped writing. You produced tons of paper. And once someone who knew you in your early life said it was almost inhuman the way you wrote and wrote and wrote. He claims that he did not see in you the love and affection for life that was so obvious in your work at a later stage in life. When he saw your *Awakenings*, he explicitly wrote you a letter saying that he was so happy because you were no longer that almost inhuman writing machine. What happened?

O.S. Certainly, since the age of twelve or fourteen I have had a need to write—experience for me must be explored further in writing. I have kept journals, I have written continuously for almost the last sixty years. Someone counted them, and I think I now have 650 journals. Most of the journals are not read by anyone. When we began we spoke of Oaxaca. This book, my *Oaxaca Journal*, is in fact largely taken from my journal of that trip.

E.P. We have the Spanish edition here.

O.S. Diary keeping is an essential activity for me. You mentioned the book, *A Leg to Stand On*. That book arose from a very detailed journal I kept of my experience. Case histories themselves are partly like journals. You referred earlier to a very good friend of mine, a gifted poet called Thom Gunn, who very sadly died earlier this year. We were both born in the same part of London in the 1930s. When we met on the West Coast in the 1960s, he was both excited and appalled by my writing. He thought I had sort of high gifts, but that I was limited by something rather cold and analytical. Anyhow, he told me he was rather pleasantly surprised, and indeed astonished, by *Awakenings* when it came out. He said that the very things that were most absent before, i.e., warmth, affection, sympathy, were now at the very heart of this book. Why, he asked? Did you fall in love? Is it through psychoanalysis? Is it through drugs you took? Or has it just been your natural evolution? This obsessed me. I carried his letter around and I find I just have to agree: it is all of the above.

E.P. We mentioned Picasso. One reason why Picasso painted such very good paintings is because he painted so much, so frequently, indeed so incessantly. Is it important to quality and excellence to work so much and to think so much? Do not you think so? Or do you think that someone different from yourself, someone who remained alone and unlike you, never wrote very much, never wrote incessantly could suddenly be inspired to come out with an incredible piece of writing?

O.S. I think sometimes that does happen, and sometimes it happens in old age. In America there was Grandma Moses, who suddenly started producing paintings in her seventies. And there have been some composers who maybe only showed their strength, discovered themselves, relatively late, even though they had composed lesser works earlier. I was both a quick developer and a slow developer, early and late. I think the chemistry was an early development, and I must have been precocious there. But it is not clear where that development led. There had to be a much slower, deeper development, which came later. It had to be slower and deeper because human beings are much more complicated than chemicals. But I think quantity is important, as well. I have mentioned that I have kept six or seven hundred diaries. Very few have been published. Very few are even worth publishing. The only journal I have published is the *Oaxaca Journal*. With my *Uncle Tungsten*, I had originally written two million words. The book was condensed into one hundred thousand words. I write many, many things. I have written hundreds of case histories, but I have tried to publish fewer than 5 percent of them. Sometimes this is because I feel it would be indiscreet or perhaps offensive to publish these things. Sometimes I cannot finish them; I do not know how to resolve this problem, and so, in the end, only a little of writing sees the light. The little that comes out comes from a big cauldron, which is working all the while, even when I am swimming. In fact, especially when I am swimming. Because I love swimming; I find it makes my mind work like nothing else does.

E.P. Our ancestors were fish. . . .

O.S. Yes. One of Thom Gunn's early books of poems was called *The Sense of Movement*. Thinking implies movement. I find that every sort of movement is crucial.

E.P. Which do you prefer, gas mantle light, or high-pressure modern bulbs?

O.S. I am very nostalgic for gas mantles with their beautiful, slightly yellowish light. A hundred years ago, it was not clear whether gas or electricity would win. Houses built at the time, like ours in London, had gas mantles as well as electricity. I should mention that there are other lights I like, like sodium light. The golden yellow sodium light first appeared in the 1940s. Some people hated them because the world becomes monochromatic under them, but I love the golden light of sodium. In fact, I have one of the few household sodium lights in existence—a friend in the lighting industry gave it to me. Sodium is among my favorite elements, and the orange-gold light is my favorite color.

— II —

Locked in the Dark

Interview with Rodolfo Llinás

We humans have an endoskeleton and crustaceans have an exoskeleton. . . . The difference is huge.
—RODOLFO LLINÁS

For some people, the brain is the most perfect organ in the universe, while for others it is simply the random by-product of a twisted, haphazard evolution.

Now we know that the mind is firmly anchored in the brain. Thanks to the investigations of Joseph Ledoux, author of *The Emotional Brain*, we can understand the difficult and often stormy synaptic communication between the limbic system, which governs our emotions, and the neocortex, which

---

Rodolfo Llinás is the Thomas and Suzanne Murphy Professor of Neuroscience and chair of the Department of Physiology & Neuroscience at the New York University School of Medicine.

regulates consciousness (see chapter 12). Corporations around the world support the fees of their top executives to attend professional seminars on "emotional intelligence." This idea is based on the premises of Daniel Goleman that it is possible to consciously improve the control of the emotions, at least compared to the state of things sixty thousand years ago.

However, no one has managed to make us forget the imperfections that surround us as the result of the actions governed by this organ, which consumes over 20 percent of the energy available to the human body. If the phrase "by their acts ye shall know them" is true, the knowledge of the brain remains uncertain. Unfortunately, the scientific and technological advances that could lead us into the age of biological control and control of the universe cannot stop senseless violence and discrimination due to race, sex, or religion, dictated by a brain that is determined to defend the genes, territories, and offspring it considers its own.

Evolutionary vestiges remain that have configured an evidently abnormal brain. The physiologist Jonathan Miller, with his Renaissance spirit, draws our attention to the alarming fact that such important organs as the heart, the liver, or the kidneys are not represented in the parliament of the brain, to which all bodily stimuli and feelings flow, simply because at the start of vertebrate evolution these organs were in different places than now. That's why, when we are in cardiac pain and our heart literally aches, the pain is felt in the arm, the neck, and the back rather than in the central-left part of the chest where the heart is now located. When a kidney stone starts to move through the urethra, the pain extends from the back of the loins to the base of the penis. Jonathan Miller says that these pains are "archaeological vestiges" of what we were. It is surprising to think that the brain, a stubborn by-product of evolution, is in control of the ship and has all the decision-making power with regard to both the management of automated processes and breathing, and discretional and conscious processes such as moving to a new location or changing partners.

No one has explained the obstacles that the brain had to overcome in order to fulfill its impossible mission, to guide moving animals so they would reach their chosen destination, quite like Rodolfo R. Llinás, who is a professor at New York University's Medical Center.

Eduardo Punset: If anything programs, it is the brain. But the brain cannot know very much because it is locked in, it is impenetrable and mysterious.

You say that our body structure is the opposite of that of crustaceans. Our soft parts are toward the outside and on the inside are the spine, skeleton, and brain. Relative to our prevertebrate ancestors it's as if we were turned inside out.

Rodolfo Llinás: Yes, but the situation is real, it's not metaphorical. That situation really happened. If you look at a crab, you see that it has a shell, it moves its claws, but if you hit it, it is hard. If you hit a cow, on the other hand, she is soft. If you push her enough, you feel that there are bones inside. Therefore, we humans have an endoskeleton and the crustaceans have an exoskeleton; that is, the skeleton on the outside and a soft interior. The difference is huge. Imagine an intelligent crab that thought, "How do I move?" It would think that the answer was mysterious, because its soft part is on the inside. This doesn't happen to us, because the soft part is on the outside and we can touch it: we touch our tendons and movement doesn't seem mysterious to us.

E.P. You said that only those who need them have brains. The plants don't have nervous systems.

R.L. Right. People think that the brain appeared suddenly. This is not true. It took 650 million years to become what it is. The process was as follows: the single-celled organisms, which remained microscopic for more than two billion years, formed a corporation, that is, an animal, because it brought them great advantages. Then, a system that could interact with bigger things was created. And thus two diametrically opposed philosophies of life arose. There is the philosophy of the plants, living beings like us that have circulation, reproduce, and die. But they lack locomotion. Thay do not actively move. If there is a fire, trees do not run off, but the monkeys in the trees do. This is the philosophy of remaining still: I cannot move and I will do the minimum necessary to survive. And then there is the other philosophy, that of movement. But in order to be able to move, you need the nervous system. So, the nervous system appeared and evolved from the need to move.

E.P. The brain really came about with those living organisms that moved?

R.L. Yes, with organisms that have intentional movement.* I want to move to

---

*Locomotion in organisms with intentional movement, some with impressive behavioral repertoires, nothing to do with the earliest evolution of will and consciousness, evolved before any amimals. While they lack any nervous system, let alone brains, many protists and prokaryotes are independent single cells, fully alive and responsive to their environment.—LM

a certain place or position. Because after all, the trees move with the wind or when the sun changes position, but animals must move in the external world and therefore they need an image, even if it is very primitive, of what they are moving toward, or away from. They may be going toward the mouth of a person or an animal that is going to eat them, for example. So, moving is very dangerous if inside you don't have at least a simple image of the outside world. That is the key.

E.P. And so the brain appeared.

R.L. Yes, the nervous system is necessary in order to predict movement.

E.P. Which organism created the first prototype of this device that was then wired up and intrinsically capable of generating images that emulate the external reality, even if they are not exact?

R.L. If we look for the animal that demonstrates this hypothesis, the missing link of how the nervous system appeared, there are some animals called sea squirts that live at the bottom of the sea. They are like a kind of bottle; they have very thin skin, sometimes of a very attractive bluish color. All they do is take in water and expel it with a filter. They simply have an entrance and an exit for water. This minimum system does not require a brain, only a primitive digestive brain that activates a simple water pump. And they don't need to know what is out there or where they are going in order to look for water, because they are surrounded by water. When they reproduce, they generate an intelligent seed, and this is the most extraordinary thing. Almost all plants generate millions of seeds, but many die or do not germinate. The seed of the sea squirts, on the other hand, the larva, like a tadpole, is able to receive light. It knows where it is, up or down. In other words, it has a light and touch sensitive system, and the possibility of very briefly understanding the outside world. The sea squirt larva moves actively but it only lives for one hour, because in an hour the battery runs out. It lacks a digestive system. It hatches with a yolk, which it absorbs as it is slowly dying. In the course of that hour it must find a place to settle to attach itself. When it finds that place, it attaches itself to it, puts its head inside, and resorbs its own brain and tail, because it no longer needs them.

E.P. There are people who, once they are settled and have found a permanent job, behave like the *Ciona intestinalis*, like the sea squirts.

R.L. And that demonstrates the relation between intelligence and the ability to move. In the course of evolution, some forms have acquired a digestive system and then they have been able to continue exploring the universe. And those forms are us, that is, the vertebrates.

E.P. Let's examine what happens when we need to predict the future a little and program it. Does that require a more complicated organism?

R.L. Of course. And then the animals develop a head. What is the head? In animals that move, the head is the part that leads. And because all new things come from the front, because when you move forward, the eyes are at the front. The hearing and all the devices to perceive the universe are where the information comes from. Then the response you can give to the outside world is much faster.

E.P. And to do that, this device must be able to represent intrinsically, on the inside, what is happening outside, even if it is in an imprecise way.

R.L. Of course. The inside body is a closed system, perforated by the senses. The nervous system has to have some idea of what is outside, based on the genetic memory and what the senses capture. With those variables, it generates an internal condition that only exists inside. Because only inside is the red of the apple mixed with its roundness, its feel, its flavor, and its smell. These things are generated by the nervous system. And we are incapable of imagining the world any other way, because we have constructed it like that. Many things we don't see: the television signals that penetrate us, the radio waves, radioactivity. We only see what is important to us. We ignore anything that is not important.

E.P. How is it possible that an apple only exists in the brain?

R.L. It's that way because if you are a fly, the apple is seen in a completely different way than if you are a horse. We make an internal image of what an apple is for us, and because it is so much ours we don't imagine it could have different values for other animals. And, without a brain, apples don't exist, we don't recognize them. What is really interesting is that the system is closed. If I sleep, I dream in great detail, with music and colors, I see things that don't exist. So I know that the pieces of things that form reality are inside me. I make things become functional. The demonstration that the system is closed is the fact that you can think of something, invent it and then make it, when that thing has never existed outside.

E.P. Give me an example.

R.L. I imagine a device with the same properties, in miniature, as the rotation of the Earth. It is called a watch and it is a very strange device, because it is a piece of metal that represents the time in a totally different world, that is, the position of the Sun and the Earth at a given moment. Nothing like it exists outside the human race. We needed a device that could think of it and make it, so our nervous system has reached such a level of development that we can turn our dreams into reality.

E.P. You mention some extremely powerful electric storms in one of your books. The device capable of constructing these distant representations that are outside, or from a dream, needs the neurons to communicate. Sometimes they come in a kind of explosion, beyond a mere rhythmic oscillation. You call it coherence, like an army that moves rhythmically, marking the beat.

R.L. If I shout, "Goal!" and bang on the table, my arm has moved and has transmitted energy to the table at the same time that my mouth has produced a sound. A functional state has been generated at the same time. I cannot sequence the parts. If you observe the inside of the brain at that moment, you see millions of cells that say "Goal!" at the same time, like the spectators in a stadium. Imagine millions of cells activating at the same time inside the head.

E.P. Consciousness. You say that life is a system—

R.L. Yes, a mitochondrion or a cell, by itself, is not alive, but the system has properties we call life. This is the principle of the bicycle. The spirit of the bicycle is in its system, not in a tire or the handlebars.

E.P. But 90 percent of the activities of this system are automatic. And the rest?

R.L. As humans, we are cerebral animals. We are born naked, with no ability to survive alone. We die almost in the form of a fetus. We are always like children. If we measure our strength within the animal kingdom, we are below zero. A hen almost runs faster than we do. Nearly any animal can beat us at running. Instead of the development of great physical strength, we cultivated a new interactive system to think and solve problems without moving. Inside the brain, many possible solutions are considered in order to develop the best one. The nervous system has the ability, but it needs to be able to breathe, digest, or move without having to think. The automatic, vegetative brain takes care of this. It lives in the brainstem and the hypothalamus. The other, the cortex, is the one that generates consciousness.

E.P. And it is the most recent, the one that grew last.

R.L. There are two large systems: the more primitive one, the one of passions, pain, what a passion is, envy, sloth, lust, eating, and feeling. This is not negotiable. You like someone or not, something gives you pleasure or not, like a reptile. The possibility of negotiating with reality only occurs with the second system, the one of the neocortex, though it is completely dominated by the passions.

— 12 —

Dialogue with Fear

Interview with Joseph Ledoux

> It is very convenient to think that we can consciously control everything, but it is also easy for the brain to act unconsciously. If it were not like that, we would be so busy calculating each one of our steps or each breath that we would not be capable of doing anything else.
>
> —Joseph Ledoux

Passion, fear, and panic—in other words the great influence of the reptilian, primitive brain in hominids—has been the focus of much of the research of Joseph Ledoux of New York University.

After our first meeting in New York, I began to suspect that the emotional

---

Joseph Ledoux is the Henry and Lucy Moses Professor of Science and Professor of Neuroscience and Psychology at New York University.

system and the conscious system are in a situation rather like that of the first computers, whose languages were incompatible. Today, Ledoux understands the emotional-conscious disconnect as imperfection in conscious access to our emotional systems. But he sowed my doubt. I think it is also incorrect to think that consciousness and emotions are hostile worlds that fail to communicate with each other. It is a waste of time to preach, "Don't take drugs," "Don't drink," "Don't run," as it is to believe that the integrated brain guarantees control of the emotions.

For millions of years, all living organisms have been evolving systems or functions to deal with what surrounds them, to obtain energy and food and be able to flee if a threat that puts their lives in danger arises. Unlike what is still taught in schools, in those processes of self-defense the unconscious plays a more important role than the conscious. In other words, the conscious control we exercise over our brain is not as important as we think.

Joseph Ledoux: It is very convenient to think that we can consciously control everything, but the brain prefers to act unconsciously. If it didn't, we would be so busy calculating each one of our steps or each breath that we wouldn't be able to do anything else. The unconscious processes are fundamental in our lives. Some of them seem trivial, like breathing or walking. For example, when we talk, we don't think about putting the verb after the subject to order the sentence, we simply do it automatically, because our brain is ready. With emotions, it is more complicated to believe that we regulate them unconsciously, but in reality they form part of the unconscious, like breathing or walking.

Eduardo Punset: You always highlight the importance of the amygdala. What part of the brain is the amygdala, which apparently controls our emotions? All you specialists agree that the amygdala is responsible for our emotions, and that the connection between the amygdala and perception in the neocortex or the overall brain is not symmetrical. In other words, the amygdala controls our brain by means of emotions or passions, but the brain can barely control the amygdala. It's terrible, don't you think?

J.L. Yes, it's true. The amygdala is related to emotions such as fear. Its function is to detect dangerous stimuli. If a bear attacks you, the amygdala detects the danger and produces a response; without you having to worry, you react to the danger. If in New York you walk along the street and a bus is about to run you over, you react and you get out of the way. The bus passes by and then you realize that you were in danger. It's a curious aspect of the functioning of the amygdala: on the one hand, it saves your life in dangerous situations, but on the other hand, it exercises control over the cerebral cortex that is greater than that which the cerebral cortex exercises over the amygdala. We know this from our own experience. When we feel anxiety, fear, or depression, we can't force our emotional brain to stop that anxiety, fear, or depression; it can only be overcome with time. When a certain emotional state affects us, hormones and other chemicals keep us immersed in that state, and it is very positive. If a bear, a snake, or any other predator attacks us, it's not a good idea to think about how our shares are doing or what we had for dinner last night. Rather we need to concentrate on what is important at that moment. As long as that situation lasts, you can't ask your brain to allow you to be free. The emotions follow their natural course.

E.P. But sometimes we would like our brain to be able to alter, alleviate, or control our emotions. Sometimes emotions like hate or love cause us serious problems and consciously there's not very much we can do.

J.L. There are two parts to an emotional reaction. The first is the reaction itself, which is an automatic response. Years ago, at the Olympic Games in Atlanta, a bomb went off, and in that instant people did not react, as was seen in the recording by CNN. People stayed still, then a couple of seconds later they started to run, and then everyone reacted and fled. We always have initial reactions and then we go from reacting emotionally to reacting consciously. It's not that we can't control our emotions, it's that we can't control our initial reaction. That is where we fail. We exercise very little control over our initial reaction, though it is the basis for subsequent control. Therefore, when we act emotionally we control the situation. Our effectiveness in controlling the situation is debatable, but we always have a certain control.

E.P. When a sudden threat arises, such as a bomb, not moving is perhaps better than running around aimlessly, right?

J.L. That's right.

E.P. Perhaps the beast will ignore you if you remain still. The emotions in that way are intelligent.

J.L. Yes. Many predators respond to movements, and though they don't have very precise vision in detail, they can detect movement. That movement might provoke reactions in the predator, and it is then that we are in danger.

E.P. So, remaining paralyzed by fear of a sudden threat has saved us in the course of the history of the species. When a lion is about to pounce on us, the best response is to stand still. You suggest that some childhood traumas, such as sexual abuse or torture, go directly to the amygdala and are registered forever, so our conscious brain cannot wipe out those memories. You say even more serious things. You suggest that these events are not recorded in the hippocampus, but rather directly in the amygdala, and there is no way of transferring those memories to the conscious brain. Does that mean that these bad memories will remain with us forever?

J.L. It's like the negative side of something good. If you are an animal that lives in the forest and you have to survive against your predators, you must remember what they are like, what sounds they make when they approach, where they are, et cetera. These details are necessary if you want to remain alive. If you are lucky in your first encounter and you manage to escape, you want to remember everything, so you don't have to learn it again. The brain has a very effective system for learning dangers and that is good, but the negative aspect is that sometimes we learn things that we don't want to remember, such as certain traumas. And in traumatic situations, the amygdala records the information but the hippocampus doesn't, as the hippocampus is very sensitive to the effects of the hormones released in response to stress. Those hormones reach the hippocampus and prevent the hippocampus from being able to memorize properly. So we have very little conscious memory about what happened. Those same hormones go to the amygdala and enable it to remember everything in detail. Faced with the same situation, there can be a very strong unconscious memory and a very weak conscious memory.

E.P. It occurs to me that in a biological future, the amygdala, as the source of passions and emotions, could progressively increase its role, while the hippocampus, which is sensitive to stress, would gradually decrease. In time, might the amygdala become stronger and the hippocampus weaker?

J.L. First of all, I want to point out that the hippocampus is not the center of consciousness; it makes certain memories accessible to our consciousness but in itself it does not represent consciousness, though the hippocampus does seem to have a more direct access to consciousness than the amygdala. In answer to your question, in the future the human brain may evolve in such a way that the amygdala grows and the hippocampus gradually shrinks due to tension. In fact, everything is possible from the evolutionary point of view. It depends on the selection pressures to which our lives are subjected and how we cope with them. Our brain evolved by adapting to lower levels of anxiety with regard to possible predators, but, on the other hand, we are in a more stressful world because of nuclear bombs and psychological and physical tensions. Our situation changes constantly and everything depends on the permanence of the changes. Evolution responds to changes that remain for a long time. And we are now in a phase of rapid change, which makes it difficult to know where our brain might be heading. The brain needs a long time to change. It seems that in the course of the evolution of the primates the connections between the amygdala and the cerebral cortex, and the connections between the cortex and the amygdala, have grown. Optimistic predictions suggest that the amygdala and the cortex will stop fighting each other and will find a balance. In the future, in the human brain emotions will not dominate consciousness and neither will consciousness dominate the emotions. Rather the emotions and consciousness will work together.

E.P. Is it true that the process that generates panic and the response to fear is very similar in all mammals?

J.L. It's true in the case of the response to fear.

E.P. Why?

J.L. Because the sensation of fear is produced in the brain cortex, in the part that thinks, and that part has evolved a great deal. Each animal has a different type of brain cortex and therefore a different capacity to feel.

E.P. So, in reality, it is impossible to know the differences between the conscious feelings of other animals and humans.

J.L. That is a philosophical question. The philosophers deal with the problem of "other minds." I can't say whether you are conscious or not, I only know that I am because I am observing my own mind. But if you were a robot, I wouldn't know it.

E.P. Your point of view on these things could change the method of curing or treating certain addictions. If you say that the impact is forever, instead of trying to convince a drug addict to change, for example, perhaps we could go directly into the amygdala and try to rub out what has been recorded.

J.L. The pharmaceutical companies will be able to act exclusively on the amygdala. Meanwhile, Valium and other drugs for the treatment of anxiety have an effect on several brain areas at the same time. Not only do they control sleep, but also sexual stimuli or fear. If we could target an effect on the amygdala exclusively, the drug would control anxiety without side effects.

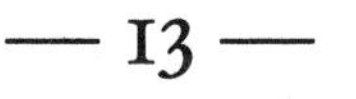

Fretful Worry

Interview with Kenneth Kendler

> The ideal in a dangerous word is to have
> a moderate level of anxiety.
> —Kenneth Kendler

At the start of the 1950s, to calm its patients, psychiatry had a sparse array of instruments consisting only of electroshock and turpentine. It was believed that electroshock disconnected the neurons and therefore the external manifestations of disturbed behavior, with the aim or the hope that when a new system of connections was randomly reestablished, the law of large numbers or Providence would cure the patient. Turpentine was injected into the arm

---

Kenneth Kendler is the Rachel Brown Banks Distinguished Professor of Psychiatry, Professor of Human Genetics, and director of the Virginia Institute for Psychiatric and Behavioral Genetics at Virginia Commonwealth University.

or the leg and produced a controlled inflammation that totally immobilized the patient, whose neurons did not disconnect at all but stubbornly clung to their original connections.

I, the son of a rural doctor, spent Sunday afternoons with my brothers with the most accessible patients of the Charitable Home for the Mentally Ill in Vilaseca de Solcina i Salou, in the province of Tarragona. I met patients like La Pedreta, who played ludo or snakes and ladders without ever letting go of a little stone that she held between her index finger and thumb, as if it were an inimitable juggling game, and which she would intermittently address, calling it by its name, "piedrecilla" (little stone).

At that time, the percentage of lost people was particularly high. It was difficult to distinguish the tramp from someone uprooted who was looking for a place to live, or from the mentally deranged. For obvious reasons, the proportion of men and women from this last group that go missing has always been higher than among the general population. To the possible desire to escape from desperation and suffering into a generic, never-identified freedom must be added the fugitives who did not want to be found. The result was that sometimes it took many years for a powerful family to find an unbalanced daughter or wife and decide to transfer her to places more coherent with her past.

On the journeys from the Charitable Home, into which one may fall by distraction, to our presumed more civilized one, our carriage was driven by a lad from Huelva, who drove with a steady, instinctive hand, as if he were the automatic pilot of a modern jumbo jet. It was a florid journey among olive fields, which years later would be cut in order to put a petrochemical plant in their place. Those journeys remained forever engraved in my young mind, and since then I have never been able to distinguish between the three aforementioned groups: mentally deranged, tramps, and fugitives reluctant to be found. Like the majority of people, it must be said, the classification of depressed patients into the category of "the ill" is a recent practice of neuroscience. The degradation of the body due to external or infectious diseases is about to be controlled. But it has taken many years for us to even contemplate the possibility of the same kind of degradation of life caused by mental dysfunction.

I didn't learn that depression was an illness in Vilaseca de Solcina i Salou. I was taught that by my friend Lewis Wolpert as he looked into thin air on one of the bridges over the river Thames, from which he had wanted to commit suicide many times during the three years he was trapped in depression. Lewis is a prestigious biologist who specialized in medical applications and popular science. His book *Malignant Sadness* has done more than any other to eradicate from so-called common sense the idea that depression is not, in reality, an illness. That little white lie is still heard frequently, despite the fact that among patients with depression 1 percent commit suicide!

If Lewis Wolpert is one of the scientists who has most contributed to the idea that depression is a serious illness, Kenneth Kendler is the scientist who has best identified the biological and genetic causes of anxiety and depression. These twin messengers pose some of the greatest threats to the health of modern citizens.

Eduardo Punset: Kenneth, it seems that anxiety is an evolutionary resource. It helps us react to danger, or the threat of an act of violence. But why do my students suffer, before an exam, the same anxiety that they would have suffered forty thousand years ago if a tiger or a lion were threatening them? What is wrong? Why can they not distinguish different degrees of threat?

Kenneth Kendler: As you say quite correctly, the world is very dangerous. Animals responded to anxiety or fear when they faced a potentially dangerous stimulus. The brain was receptive to the stimuli or dangers that surround us, because that predisposes us to act against danger. But excessive worry about danger takes up too much time and energy. It reduces survival. What has happened in evolution? The genes have developed to predispose individuals to certain levels of anxiety, and, ideally, in a dangerous world, we should have a moderate level of anxiety. Obviously, to not be worried by danger is not very intelligent in evolutionary terms.

E.P. It seems that our genes have us well prepared for the environment in which we lived forty thousand years ago. We are well prepared to fight against threats of forty thousand years ago, but we are pretty badly trained for our modern environment in which the threats have changed.

K.K. Exactly. It is fascinating for those of us who work in this field. Why do people see their doctors because they fear spiders, snakes, rats, or lightning, and, on the other hand, no one has an irrational fear of firearms, cars, or electrical appliances, which really hurt children? Human beings seem predisposed to develop fears toward stimuli that were dangerous when we were evolving. Evolution has not had time to come to terms with the fact that today firearms or electrical appliances are more dangerous than a snake or a spider.

E.P. Kenneth, the English neurologist Susan Greenfield maintains the thesis that depression and anxiety are the most human mental illnesses that exist. She maintains that what distinguishes us is a brain with too much capacity to reflect on our past, present, and future, which causes trouble. People who suffer from depression are sick of their own minds, unlike schizophrenics or children who see fantasy worlds and hallucinate. What do you think? Would genetic factors be less in cases of depression and anxiety?

K.K. It's complex. Depression in particular is more a disorder of the higher cognitive and emotional centers. I don't think that there is a good model for depression in rodents. There is evidence that the large primates—the chimpanzees, and, of course, humans—develop depressive syndrome that affects self-esteem and self-evaluation. One of the causes of depression is a negative evaluation of oneself. The forms of anxiety are, I think, rather different. Rodents present many characteristics of anxiety. In the past ten years the basic neurological routes that intervene in the behavior of fear in rodents, primates, and humans have been shown to be essentially the same. With new techniques to reproduce images and, in particular, with noninvasive magnetic resonance spectra, most work focuses on the neurological structure called the amygdala. In humans, chimpanzees, and rodents, this part of the brain plays the role of central connection in development of anxiety. I think anxiety is phylogenetically older than depression.

E.P. You mean we share anxiety with rodents and primates and that this has been genetically demonstrated? When we speak about genetics and pharmacogenomics, drugs used in genetic therapy, we think we understand better. But the patient doesn't understand very much. What do you think?

K.K. First, I'll answer you in general. It is important when talking about complex phenomena like anxiety that we do not think of an either/or situation—either the genes or the environment. We tend to think that if genes are involved, it is like their determination of the color of our eyes or our blood group. These illnesses are very complex probabilistic phenomena. In the disorders of anxiety, some 30 or 40 percent of the differences in predisposition among people is probably due to genetic differences. This is similar to what happens with blood pressure. We know that genes we inherit from our parents influence the level of blood pressure. But diet, smoking, stress, et cetera, the experiences of our environment, are all important. People ought not to think that when we talk about genes we can assume that the environment plays no role. In disorders of anxiety it influences a lot. Many phobias arise owing to very traumatic events. I had a young patient very afraid of heights. His father had a wicked temper. When my patient was three or four years old, he lived on the sixth floor. When his father was angry, he held him out of the window by his ankles until he calmed down. The child was afraid of heights even if he never had any genetic predisposition. He had suffered a sufficiently traumatic experience. A complex relation during development exists between genetic predisposition and the conditions of the environment that alter that predisposition to anxiety. To what extent do patients know how to respond to this illness? It depends enormously on the type of psychiatric attention they receive. For illnesses related to anxiety, we need to know that not only do we have pharmacological treatments but also behavioral treatments that are very efficient. Phobias, the most widespread form of anxiety, can be treated rapidly and efficiently with behavioral therapy. He who is afraid of flying is not treated with drugs, but rather there are numerous behavior therapies. Now, with the world of virtual reality, there are people who use 3-D glasses for virtual-exposure therapy, instead of getting on a plane as they did before.

E.P. How long will it be before we have drugs without side effects that act only on specific mutations or gene expressions? Twenty or fifty years?

K.K. You ask me for prophesy. That is very risky. To me the most exciting news in molecular developments for psychiatric disorders relates to understanding, at the basic scientific level, what are the anomalies in these illnesses. In psychiatry we are in a less advanced position than our colleagues who work on the genetics of cardiovascular disease or diabetes. We understand at a very basic level which enzymes intervene or what transports cholesterol from one place to another. And, though we have made great progress in psychopharmacology in the treatment of these disorders, at the basic biological level we still don't know what abnormalities they present. When we are capable of identifying specific genes, I'm sure we will find there are many; it is very improbable that a single gene is responsible for much of the variability of these illnesses. The great advance will be to have opened the door to biology for the study of these illnesses. I am old enough to remember that twenty-five years ago we took urine samples or lumbar punctures to obtain spinal-cord samples to seek neurochemical anomalies. This was condemned to failure. In these illnesses, perhaps hundreds or a few thousand of the thousands of millions of nerve cells are involved. Those measurements were far too crude.

E.P. As a physicist interviewee said, we're the wrong size [see chapter 33]. We humans are too small to see galaxies and too large to observe the tiny real world of atoms, molecules, and cells. This diminutive life has meant something to you. You hold experimentation in such high esteem that you believe that even psychiatry has a biochemical basis.

K.K. The question you ask about the relation between the mind and the brain is really fascinating. As a psychiatrist, I think I am in the most privileged position to understand how mind and body relate. Of all the medical disciplines, psychiatry is the one that tries to relate the mind with the brain. Let's take, for example, a typical clinical situation: I am treating a young woman who has suffered an episode of depression. We talk about how she feels and about a marital problem related to her difficult relationship with her father. We talk about profound questions of her mental world, her identity, and nonetheless, in the course of the visit I have to disconnect and think about what is happening in her neurological system, how it affects the hippocampus and the amygdala, and whether I give her the right dose of medication to regulate the receptors. Then I must disconnect again and return to the world of the mind. So in our daily work, we psychiatrists always think about this inextricable interrelation between mind and brain.

E.P. Marvelous. While you were speaking, I wondered whether, in the future, when we are capable of moving from the mind to the body, if people will continue to see their doctors in search of happiness. Once we combat anxiety, will we come to you for something more?

K.K. There are extreme biological reductionists in psychiatry who believe that when we can understand all the molecules, we will only need to hand out pills. I think this future is terrible. A fundamental aspect of the human condition is that we are animals who tell stories. We need to understand our own stories. It wouldn't be right if Mrs. Jones came to surgery with her marital problems, which are a reflection of her problem with men like her father, and I said, "Mrs. Jones, you are suffering from low regulation of the neurological system; take these pills and you'll feel better." That would not suffice. We are humans and need to understand our experience. I don't think psychiatry will change much; I don't think it should. One great challenge is to be capable of thought about the meaning of our lives, as well as the physical basis that operates in the brain.

E.P. I have a specific question related to Catalunya, the corner of Spain where I was born. Two scientists, Antoni Bulbena and Xavier Estivill, suggested the existence of a direct correlation between the slackness of the joints of the body and anxiety. This was interpreted as one of the proofs for the genetic basis of anxiety. Are you familiar with this?

K.K. Yes. I think it is an intriguing discovery that deserves further investigation. We don't yet understand the basic science of anxiety. I'm not saying it's impossible for it to be related to the gene that controls the slackness of the tendons, but the study has not been sufficiently reproduced. We know from the history of genetic psychiatry that many initial findings do not pass the reproduction test. So the interest and enthusiasm are justified, but other groups must study this before we can accept it as true.

E.P. Kenneth, do you agree that virtually all diseases that come from outside have now been controlled and now the challenge is to control those that come from inside us?

K.K. I do agree in general. An exception is the AIDS epidemic. Infectious diseases, in the West, are practically controlled, unlike chronic diseases like diabetes and hypertension. Psychiatric disorders are one of the main causes of illness and mortality. The World Health Organization did a study of their development up to the year 2020, listing the illnesses according to their impact on health levels. It predicted that by that date disorders like depression will have overtaken cardiovascular disease as the main cause of illness and death. Alcoholism and anxiety will also be among the top ten. So if we think about the general causes of human suffering, the degradation of life and mortality, nervous and behavioral syndromes will become increasingly frequent. I think intensive research efforts are justified to understand and alleviate this suffering as much as possible.

PART TWO

ANIMAL BODY-MIND

Cyclicity and Sociality

INTRODUCTION

The more we know about other forms of life on Earth the more we are impressed with their genius. What is most striking is the cyclical nature of the living process. Rhythms abound. Day becomes night becomes day. Spring becomes summer becomes autumn becomes winter becomes spring. Seeds sprout, spores germinate, eggs are laid and then hatch, babies and puppies are born, trees flower, skins are shed, ancient bodies pass away. Girls bud teats and boys sprout beards. These changes in all of nature are keyed to clues in the sky, Earth in its rounds about an elliptical orbit, the Moon and the tides and the almighty Sun. Inevitable patterns of lifetimes respond to water and electricity, lightning aloft, the charged particles of the Earth's magnetic field in the air, the radioactivity in the ground. Without environment there is no life at all. Independence, as Vladimir I. Vernadsky* (1883–1945), is a political, not a scientific, idea, for independence of living organisms from the biosphere at the Earth's surface, as all cosmonauts and astronauts know, means certain death. Although the cycles of birth, growth, reproduction, and death are species specific, they are sensitive to environmental clues. The patterns of life for any live

*Vladimir Ivanovich Vernadsky, in his classic book *The Biosphere*, showed how all life is connected by energy and material flow. The main source of energy with the biota (sum of microbial, animal, and plant life) insures continuous cycling of the chemical elements mandated by the living process [carbon (C), hydrogen (H), nitrogen (N), oxygen (O), sulfur (S), phosphorus (P)]. His work is known throughout the literate world and he is considered the founder of biogeochemisty. Vernadsky wrote that "life is a geological force," the most important geological force, and Lapo (1987) applied Vernadsky's thought for the reading public in his book. The relation between the work of James E. Lovelock, who never knew of Vernadsky's work when he wrote his Gaia papers and books, and Vernadsky's on science is well described by Westbroek (1991). A recent review of "Whole-Earth Science" is in Harding (2006).

being cannot be changed, nor may they be rushed. They can only be disrupted, perverted, warped, or halted. The potential for any twig on the tree of life to snap off and die is present at all times, in all places, for all forms of life. The drum beats on, life dances, the survival imperative is not negotiable, and change, not stasis, is the rule. People evolved late in the game of life. As newcomers we have a lot to learn from our predecessors. We talk, therefore we lie. We dream, therefore we deceive ourselves and our brethren. We sing and dance but our memories, and indeed even our accurate-as-possible science draws pictures far more similar to the stick figures of kindergarteners than to the photographs of Antonio Vizcaino or Ansel Adams.*

As for sociality, social lives, even the bacteria have them. Cyanobacteria will sacrifice themselves and their reproductive prerogative to incessantly provide needed nitrogen for their immediate relatives. They, unlike any animal or plant, can take their nitrogen necessity directly from the air. Slime mold amoebae will die so that their clones will be lifted on "sporophores," stalks that grow from wet ground to dry air and form balls, or bags, full of propagules, cells in packages on high that are released to scatter the spawn [see chapter 30]. We did not invent mother's clubs or orgies or flocks flying in formation. Social behaviors—attraction and avoidance, crowd survival and scapegoat deaths, electrical pulses with meaning, self-sacrifice, cannibalism, warfare, click-sound communication, and many more—are behaviors as old as life itself.

*de la Macorra and Vizciano, 2006; Adams et al. 1987.

Drumbeats

Interview with Steven Strogatz

> We have assumed that all systems require leadership, that we always need an internal, centralized command, and that's not true. And many systems would work better if we allowed them to organize themselves.
>
> —Steven Strogatz

Steven Strogatz is a modern scientist who could give us most direct answers, indeed he could satisfy the man on the street, if so many in the scientific community were not so prone to confuse the less informed! Some scientists, but not Strogatz among them, seem to think that the questions of real interest to people should never be asked. Behind such enigmatic phrases as "systems of self-government that execute phase transitions"—the usual academic jargon—

Steven Strogatz is Professor of Theoretical and Applied Mechanics at Cornell University.

lie the search for why the menstrual cycles of a group of women working in the same office tend to synchronize, the shock that takes place in a cell that suddenly decides to become cancerous, the crash of the Stock Exchange, the appearance of consciousness in the brain, the synchronized beating of the cells of the heart, or the determination of the molecules of vodka left too long in the freezer to remain locked together in the form of ice, refusing to flow through the neck of the bottle into the glass. Strogatz tries to answer these conundrums in plain language, even if he disappoints specialists.

Eduardo Punset: Do we really know what happens in phase transitions, synchronized menstrual cycles, and the transformation of cancer cells?

Steven Strogatz: To me they are some of the most interesting questions in current science. And in almost no case do we completely understand them. To give a few more examples, I would mention the stability of ecological systems composed of many species, and the organisms that feed on each other and interact and cooperate. We start to learn certain things: we have found similarities in all these systems because, for example, they are all composed of many individuals that interact following simple rules, but from the combination of all the individuals an incredible complexity arises.

E.P. Let's talk about a miracle that you mention several times in your book *Sync: The Emerging Science of Spontaneous Order*—the fireflies of Thailand. There we have several hundred fireflies that suddenly emit light all at the same time, in the same instant, though we don't know how they communicate with each other. Or is it true that it's a question of sexual competition, and that it is the males that try to surprise the females by being the first to emit flashes of light? How on earth do they all do it together?

S.S. Yes, I think it is one of the few cases relatively well understood. It is only male fireflies that take part in this spectacular phenomenon. This phenomenon occurs in Thailand where, along the banks of the rivers that flow into the sea, there are thousands of male fireflies in the leaves of the trees, and at night, when the sun goes down, they start to see each other as they stop competing against the sun. And at first there is absolutely no order; they emit light randomly, without being coordinated, like an audience applauding at the end of a performance in the United States. Here, we are so independent that we are completely disorganized and don't even try to do otherwise. In countries like Hungary and Romania, and perhaps Spain, the cultural expectation is for everyone to applaud at the same time. But, in the end, we are intelligent human beings and we can do it; the fantastic thing is that the fireflies, which are not particularly intelligent, can produce a similar visual version. What happens with the fireflies is that at first they are a bit disorganized and then, in small groups of two, three or four, they start to emit light and, after one hour—it's not instantaneous—a mile of the bank of the river is covered in trees with these insects emitting light at the same time. And we understand it because a firefly not only emits light but it can also see. So its nervous system adjusts unconsciously: when it receives a flash it can increase or delay its internal flash controller. So this starts to flash in harmony with their companions. We don't really know why. Do you want to talk about motivations?

E.P. It would be very interesting to know why. But, finally, they all applaud at the same time, like audiences in Hungary. Another fascinating example inside us is the biological clock, these circadian reflexes that make us wake up or go to sleep more or less at the same time. Is this circadian cycle in the brain or in the cells?

S.S. It's a fantastic question. We have circadian cycles of which we are all aware. It is possible to wake up every day at the same time without the use of an alarm clock, because something in the body tells us the time. The question is, where is it? I would say in each cell of the body. Every cell has biochemical cycles that keep it synchronized with the world around us, twenty-four hours a day. There is a master clock, as you indicated, which is the brain, in a very interesting location. We have two eyes, which are in contact with the world outside, with life. Inside the eyes, we have the optical nerves that connect with the brain and that come together to make that connection with the brain. At this intersection, where the two nerves come together, there is a group of thousands of cells, deep inside the brain, in the hypothalamus, and these cells are called the suprachiasmatic nucleus. I'll explain the etymology of that word: it comes from the Latin *supra*, which means "over," and *chiasma*, the optic chiasma, which is where the optic nerves cross. Each cell is like a little electronic clock that is kept in something like a Petri dish with its own nutrients, and each cell has a rhythm of voltage that rises and falls in a cycle of twenty-four hours. It does not need the rest of the brain or the body. They are little clocks that are inside our head.

E.P. It's incredible, but then why do babies—all parents have experienced this—not know when to wake up and when to go to sleep?

S.S. I don't know. Many parents realize that after three months babies start to synchronize with the cycle of day and night. And I'm not sure what happens in those first three months, perhaps the cells are not able to communicate with each other properly; it's possible that the synapses are still developing in the brain. We know that at that stage of growth the child's brain is developing. And not only that, but also some of the connections between the neurons are being cut. There is the very interesting process, which takes place during development, of pruning the neuron connections. Babies have too many connections and have to sever the unneeded ones. I don't know the right answer, and perhaps there isn't one. But something happens in the baby's brain that enables it to synchronize after several months of life, although in reality we don't know why.

E.P. What you've explained seems reasonable. It's likely that the cells need a certain time to adjust this system of communications among each other. In the circadian cycles, there are the famous zones which you call zombie zones. I lived for three years in Haiti and I know some things about zombies, but I didn't know about these zombie zones in our circadian cycles. Steve, correct me if I'm wrong, but between three and five o'clock in the morning we should not be awake, because if we are the neurons suffer and if we insist on keeping them alert and working, disasters can happen, like Chernobyl, the Exxon *Valdez*, or the Bhopal tragedy in India. Those accidents happened during the zombie time, didn't they?

S.S. Yes. As you said, the zombie zone is a dangerous time. It's a particularly delicate period of time in the circadian cycle, when many biological functions are at their worst. So, our cycle of wakefulness reaches its lowest point around that time. The reason we call it the zombie zone is because when you have to be awake all night you're aware of it: you feel tired, your eyes are sore, and you feel lethargic. It's due to the function of the biological clock in the brain. And we have all experienced that if we remain without sleeping for a little longer and it starts to get light, we start to wake up, even if we can't see the light. If you are a person who works with computers and you stay up one night to finish a project, you don't have to see the light that starts coming in, you feel that you are waking up, as if you were getting up from a second siesta. This second siesta is the product of our biology, and it is when the body is starting to wake up. There is a hormone called cortisol, which is secreted by the adrenal gland, the body's stress hormone. It prepares us for the activities of the day, for the battles we have to fight during the day. At the end of this zombie zone, this hormone starts to be secreted. The temperature of the body goes up from the minimum, many functions of the body start to connect, and the body begins to wake up. But, as I said, before that moment the basic functions are at their lowest point. If you are a pilot and fly a plane during that time, you might not be sufficiently adjusted, or if you are the controller of a nuclear power station and you start to work on the night shift—this is the moment when the most serious accidents can occur.

E.P. What advice could we give people who have to work in the zombie zone, around three o'clock in the morning, and young people who go out clubbing and are surprised that at six o'clock in the morning they start to wake up again? If they go to bed at six or seven o'clock in the morning, thinking they will be able to sleep for eight hours, they're wrong because after a few hours they wake up because of the cortisol you talked about, right?

S.S. Yes. If you are awake all night and when you get home you want to sleep, it will be difficult. You might think it's because of the excitement of partying. Perhaps in part it is, but also the body's internal alarm clock is starting to call: the temperature of the body rises, cortisol is secreted, and other hormones start to be produced. In other words, the brain begins to wake up. It's a big problem for people who work at night, because when they go home, after having worked, they need to sleep because it's the only time they have to do so, but often they can't because of the noise of the traffic, the children, or the light entering the bedroom. It's assumed that it's these factors that prevent them from sleeping, but even in ideal conditions the body wouldn't be able to sleep at that time. I want to give a very different piece of advice to people who travel through different time zones and suffer from jet lag. Each person feels it in a different way: some people say they never suffer from jet lag, because they sleep on the plane and when they arrive they are rested. These people believe that jet lag is a phenomenon of lack of sleep, but that's not true. Jet lag occurs when the body's circadian pacemaker, the internal clock, has remained in the previous time zone. If I travel to Spain from the United States, my zombie zone would happen at a time when I am doing something, for example, in the middle of a meeting, and that's a problem. That is what we call jet lag.

E.P. Some people say it's easier if you take some sort of drug.

S.S. There are people who swear it helps, but I don't know. My impression is that this field of science is just emerging. I would take all kinds of precautions before taking any drugs, such as melatonin, for example, which is the most popular one. The studies that have been carried out into naturally produced melatonin demonstrate that the brain secretes it in tiny amounts: a trillionth of a gram of melatonin will be active and will affect the brain. It is a brain hormone and I would take it with a great deal of caution. Without knowing what it's doing, it doesn't make any sense to me. I think it is more important to expose yourself to the light of day. The sunlight is what resets the biological clock, so if I had to make that trip to Spain, what I would do is immediately adapt to Spanish time. I would go out to have breakfast and take a stroll around the park in the morning. And at night, I wouldn't stay up too long, even if I didn't feel tired. I would go to bed at a reasonable time in that time zone. So, to reset the internal clock the best thing to do is to expose yourself to sunlight. This gives good results very quickly.

E.P. Better than melatonin. There is another fantastic case, which you call the sinoatrial node of our heart, the pacemaker that keeps the heartbeats regular. It seems that this is also a process localized in a tiny area: it's about ten thousand neurons that come to an agreement about when they have to beat. How does it work? How do they do it? Isn't this a miracle? Do we know anything?

S.S. Yes, it's a little miracle about which we know very little. It is the miracle that keeps us alive. Each beat of your heart is the result of the work of these cells, whose function is to send an electric signal to the rest of the heart, to the ventricles, which are the pump chambers of the heart. These pacemaker cells play a very important role. If they don't drive the heart with perfect coordination, the result is a conflicting signal, the heart wouldn't know what to do, the blood would not be pumped and you would quickly die. So the design of nature is marvelous. If there were only one pacemaker cell, a single cell that was the leader, it would apparently be a simple solution, but it would not be a solid solution because that cell might function badly and die and then you would die. Nature has created this kind of democratic solution: There are ten thousand cells responsible for telling the rest of the heart when it has to beat; but, as you said, there is no leader. They need, through a democratic or self-governing process, to reach a consensus on when to give the order. They do it with each beat of your heart, for around three billion beats in a lifetime. They never make a mistake. How do they do it? They are in constant communication with each other when they give the order. When the voltage rises and falls they send electric currents, ions, through electrical connections called gap junctions, which connect all the cells together. So, through this constant exchange of electricity, they are in communication, which is sufficient to allow the slower ones to hurry up and the fastest ones to reduce their speed. They regulate themselves automatically. If one dies, it's not a problem because the group continues to work perfectly.

E.P. The other 9,999 cells would continue to do the work. Another miracle: Is it true that if five or six women were locked in a room they would menstruate at the same time? It seems incredible; how does one woman communicate with the others? Through chemical substances or through the mind? What is it that makes women synchronize their menstrual cycles?

S.S. Tests show that through chemical substances not yet identified, probably a chemical product in the sweat. These tests come from a study carried out at the end of the 1970s. A woman called Genoveva noticed that every summer, when she went back home during the university holidays, her sisters' menstruations synchronized with hers. The doctors who studied it thought it was because—I can't say because of her smell, because she didn't give off any smell, but there was something that she produced chemically, and that was transmitted to her sisters. To demonstrate it, they put cotton swabs on her armpits and every day took these swabs that contained sweat, ground them up and mixed them with alcohol. They called it "essence of Genoveva," almost like a perfume. Then, they took little samples of this "essence of Genoveva" and they put it above the upper lip, below the nose, of women who lived thousands of miles away and who didn't know her, had no contact with her. These women were capable of smelling the chemical products present in the sweat secretion. And the incredible thing is that in a few months their menstrual cycles were perfectly synchronized without ever having met her. So, in her sweat there was something that told them about her menstrual cycle and it was sufficiently powerful to synchronize women whom she had never seen.

E.P. It seems incredible. This is an obligatory question: what about fashions or epidemics? Do epidemics caused by viruses, which suddenly spread, also follow these laws of synchronization?

S.S. There is a certain similarity in the spread of a virus in an epidemic, and, let's say, the spread of the models of the synchronization time of the fireflies or the cells of the heart, but there are also differences. I think that the propagation of the virus is done in a more sequenced way. First, it starts in a group that is infected and then the infection gradually grows, as if it were a wave. There is not the same simultaneity as in the heart, when all the cells emit a signal at the same time. This is more like a wave of propagation but, of course, it is another example of cell organization. Synchrony is the simplest thing, when everything happens at the same time; propagation in waves is a more complicated model in the self-organization of a group. I think that this is the analogy and the difference with fashions or epidemics.

E.P. Let's see. A question I ask not so much to you as to the wind. We are surrounded by computers connected in networks, telephone links, the cells connected by chemical products; we have companies and consumers connected by the market. Steve, I am still very interested in management subjects and when I analyze your discipline, my immediate question is the following: don't you think that in the world around us there are too many systems that are not self-organized? What I mean is, isn't there excessive interference, of a bureaucratic nature, with systems which otherwise would work well on their own? We can see it in corporations or large institutions, where there are dozens of people supervising what is done, when, in reality, what they do is to interfere with people's innovation and peace of mind. They behave as if they were not self-organized systems.

S.S. A very intriguing question; your idea is very attractive. Yet it seems to me that there are systems that need to have governors, leaders. But perhaps your question refers to the fact that we have assumed that all systems need leadership, that we always need a centralized internal command. That is not true. And, as you say, many systems would work better if we allowed them to organize themselves. Adam Smith said that about the economy many years ago, that the invisible hand would take charge of the economy if you just got out of the way. But, for me, this is not a scientific comment, but rather a very interesting philosophical thought.

The truth is we don't know enough economy from a scientific point of view to know whether experience demonstrates that capitalism is better—at least, in the countries with free markets we tend to think it is better, but scientifically I still see it as an open question. That is, we can't carry out controlled experiments with the economy, and we don't have a good mathematical understanding. We still have very intelligent people who try, who are very well dressed and spend their lives discussing whether we should lower taxes and if that would help the growth of the economy. This happens constantly in the United States with George W. Bush, who says that lowering taxes would promote growth. Others say that it would cause such a great deficit that in the long term it would be worse. And in reality nobody knows, because there are good arguments to defend both positions.

I think that it is a great opportunity for science to consider all the options, analyze how these complex interconnected systems work in reality. That is why we study things that might seem ridiculous, such as fireflies, but which are sufficiently simple that we are able to make some progress on a problem that essentially has the same characteristics as a truly important problem like the economy, global warming, or the politics of many countries, which are interrelated. These are the things that we really want to understand, and by study of things that seem irrelevant, such as the fireflies, we can have a vision from inside. This first step helps us to understand these very complex systems that have a great impact on human affairs.

E.P. One last question, Steve: you, as a great specialist—I don't know anyone who has studied graphics, networks, and synchronization in such great depth—when you feel a slight discomfort in the stomach that indicates that it is digesting the food, which it does automatically, because it is a self-organized process, do you feel more secure about the signals of this automatic process than about the decisions you have to take consciously, because the process is not automated? I'm referring to decisions like accepting a job, getting married, or traveling. Do you feel better with something that is self-organized or with something that is not? Or is it all the same to you?

S.S. I love your question. If I've understood you properly, your idea is that instinct and intuition are more self-organized processes than highly elaborate, logical conscious thoughts. Through the important decisions in my life, I have learned that it is better to trust your heart. It is a romantic idea, but also very practical. When we make decisions with our hearts or our stomachs, they are always better, I don't know why. Perhaps it's because when we make these decisions it is because there is a greater degree of . . . I'm lost, I can't find the words. I think that everybody knows that this is true, that when we make decisions we have to listen to our heart, because it has all the important information. When you think of the possibility of giving up on an idea, or when deciding between advantages and disadvantages, or when there are good reasons on both sides, normally the heart knows. I don't know. This is another good question for science. Why are the heart and the stomach better at making decisions than the brain?

— 15 —

Survival, Not Truth, Is Imperative

Interview with Richard Gregory

The brain makes many suppositions. . . . The images received by the eye do not correspond.
—Richard Gregory

Out of the window of the office of physicist Eugene Chudnovsky at Herbert Lehman College, Bronx, New York, the only thing that can be seen is the bare structure of an immense birch tree. In the middle of winter, the trees are like a scene submerged in fog seen through an X-ray screen. My question is pertinent: "Does the universe really exist?" [See chapter 31].

The reader will find the answer in this chapter. This primordial question, which probably has been on every person's lips since we learned that the first arthropods appeared more than four hundred million years ago in the

---

Richard Gregory is Emeritus Professor of Neuropsychology at the University of Bristol.

Paleozoic era, presupposes another, previous question from which all reflection derives: "Can we trust our brains? Do the cells work in concert?"

In different species, cells discover that some of their neighbor cells are not to be relied on. In animal tissues certain cells break the consensus. Unexpectedly and without warning or apparent reason, they begin to behave on their own. They transgress the pulses and rhythms of the tissue. The dissident cells ignore the others. Suddenly they begin to leave their positions and ceaselessly multiply. They cause cancer by metastasis: loss of differentiation and failure to remain in place until later when the tissue is replaced by younger undifferentiated cells that grow by division. They move around. The different genetic balance of the entire animal body breaks. Eventually death ensues. Other times, certain neuronal dysfunctions generate strange, dangerous behaviors, particularly in mammals and, above all, in the social primates. Evolution has made it clear that, in certain circumstances, internal pathologies arise that prevent the other cells from trusting their close relatives.

Another, more recent, question is just how confident in their own brains social primates can be under normal circumstances. In both its prescientific stage and in the midst of the explosion in neuroscience, mankind assumes that the brain is the most sophisticated organ in the universe. Eminent physicists like Alfredo Tiemblo dedicated many of their reflections to deciphering the mystery of why the universe, composed of atoms, has recourse to the brain to think about its origins, scrutinize its purpose, or, simply, note its absurdity.

The neuroscientist Richard Gregory has a far from reverential attitude toward the king of organs—which consumes almost a quarter of the energy generated by the human body. The workshop set up in his house in the well-to-do, hilly part of Bristol, England, offers not only the best view over the city and the sea at its feet, but also of space. Gregory has every imaginable device to play hide and seek with his brain. It is an all-out war fought over decades to penetrate, by means of experimentation, into the camera obscura from which the brain scans—by whatever intricacies—the light and shadow of the exterior.

Eduardo Punset: Is there anything more important than a correct perception of the world in which we live? We spend a pittance and very little effort on investigation of accuracy of the images we have of objects around us. How do we perceive?

Richard Gregory: You are absolutely correct. The brain makes many suppositions and obtains small images from the eyes, but it's not enough. The images received by the eye do not correspond at all to the objects it looks at. They are not identical. We look at a table and sense that it is solid and strong. We infer that we can safely place things on it. The brain has to guess that it is solid and strong. The brain imagines a real object based on an image from the eye the size of a postage stamp. It is incredible that we can go from a little image to the sense of the reality of the world. Of course, this doesn't always work. More research should be done. It should be taught in schools as it is a vital question for children.

E.P. In schools they only give answers. They don't teach the children to ask questions. I find it hard to imagine schoolchildren wondering whether the other students are like they see them or, as Newton said, how we go from the perception of the object to the glory of colors. Children, and the entire population, are denied the custom of questioning reality. The majority of people are convinced that the eyes send a faithful image of objects to the brain. You explain why this is false in your books: the eyes only send codified images in the form of electrical impulses. Who knows how the brain deciphers them!

R.Gr. Right. Of course, inside the head is complete darkness with only small electrical impulses that come from the senses, which the brain decodes. And the signal is very different from the reality we observe.

E.P. You remind me of a well-known neurophysiologist from New York, Rodolfo Llinás [see chapter 11]. Do you know him? He says something very similar. He says that, unlike the crustaceans, our skeleton and our brain are on the inside, whereas flesh is on the outside. The brain is completely enclosed and in the dark and, like you, Llinás does not think that the brain is a faithful interpreter of what is going on outside, because it receives very little, very vague, information.

R.Gr. I entirely agree. Before, you referred to colors. If we focus on colors, we will see, for example, that the carpet in my office is red, that is, it has a light wavelength that stimulates the eye generating the color red in the brain. If neither sight nor the brain existed, there would be no color in the universe. Color is created in the brain and is projected, psychologically, onto the object. That is why we differentiate red from green, black from white. Everything is in the brain, not in the outside world.

E.P. What would artists say if you told them that colors are not in the world, but rather in the brain?

R.Gr. Well, not long ago I had a public debate with a great artist at the Royal Academy in London and he was very indignant, because he is a great colorist and paints very beautiful pictures. Of course, he was convinced that the color was in the work. And I told him that wasn't true. Another great controversy with artists is perspective, discovered recently—in the Renaissance—though the brain has known about it for millions of years, since sight has existed. When an object moves away from us, the image becomes smaller in the eye in such a way that when the object is twice as far away, the size of the image in the eye is reduced by half. That is why parallel lines converge in the distance, as happens with railway tracks. It is astonishing that artists did not discover perspective before Filippo Brunelleschi, the great architect who built the dome of the Florence cathedral. It was he who noted that the convergence of lines gives depth and perspective. He applied it to his designs and since then almost all artists have used the technique. But before Brunelleschi people didn't understand perspective, though the brain had been using it for five hundred million years.

E.P. Something that most intrigues me is the fact that we cannot, literally, observe our own facial expressions. Without the ability to see our faces, we manage to relate private emotions with public expressions in other people. As you say in one of your wonderful books, we have mirrors. I quote your words: "If it weren't for mirrors, which have helped us to learn something about this fascinating mystery, we would be incapable of recognizing our own photos." Is that true?

R.Gr. It is what I believe. First of all, a baby or a child recognizes itself in the mirror thanks to movements. The child moves and his or her image in the mirror also moves. Sooner or later he or she reaches the conclusion that it is him- or herself reflected in the mirror. If mirrors did not exist, we would never know what we look like. And, as you say, we read other people's minds from the facial expressions that they themselves cannot see on their own faces. It's extraordinary.

E.P. It must be very difficult! And, nonetheless, children unravel this mystery quite early.

R.Gr. Children recognize themselves shortly before they are one year old, at around ten months. Before that age, they don't know that the figure reflected in the mirror is them. Then they learn it through the movements of their bodies, from their gestures. If children only see themselves in fixed photographic images and not in mirrors, I think they would not recognize themselves at all.

E.P. And other animals? Chimpanzees also recognize themselves?

R.Gr. Chimpanzees are the only other ones. No other animals recognize themselves in a mirror.

E.P. I don't understand how children learn to recognize themselves so fast.

R.Gr. Children learn by touching. To recognize the objects around them, they need to touch them and see them simultaneously. Therefore, sight is educated from touch, which is a primary sense. When we look at an object, though the initial image is only optical, we also see, in a way, nonoptical properties: we can tell whether it's hard or soft, if it's edible, if it's hot or cold. This all happens through association with the other senses, because all the senses are connected.

E.P. It's strange that you should mention touch! In a way, blindness is touch without sight and the mirror is sight without touch. You suggest that in a world full of mirrors, but without touch, we would never learn.

R.Gr. Yes, that's what I believe. Perhaps we would see models, but not concrete objects, and things would lose their meaning. Perception is acquired by touching and seeing things at the same time. We should not forget this in science museums. It is very important to be able to touch the objects. If everything is in glass cases, it is very difficult to know what they are. We need to interact, explore through touch, as babies do constantly. First, they put things in their mouths, and then they touch them. Their eyes come into play much later.

Richard Gregory left his workshop for a moment to look for some pieces of wood with which to make an impossible figure. His idea that perception is often the subproduct of a combination of all five senses reminded me of the anecdote of a student who asked me what synaesthesia was. I told her, "Everybody knows the word anesthesia, it means no sensation. Well, synaesthesia means combined sensations. Synaesthesics see the number five, for example, as being yellow." "No, the number five is green," she immediately replied. She was synaesthesic without knowing it, like many creators. Richard Feynman referred to one of his equations in this way: "with j's of a luminous brown, n's of a light bluish violet and dark brown x's." And the writer Vladimir Nabokov told his mother, while he was playing with colored cubes for learning letters, "They are all wrong: this cube has the A in red, and the A is blue." His mother, who was also synaesthesic, understood him perfectly. Another, more complex case is that of the musician Olivier Messiaen. Generally, the synaesthesic can listen to music and see a color at the same time. But the process does not go in both directions: a given color does not make them hear sounds. Olivier Messiaen, on the other hand, sees colors when he reads a score, and colors suggest music to him.

According to Richard Cytowic, synaesthesia occurs due to an anomalous functioning of the limbic system, responsible for the emotions, memory, and attention, in which the key is the hypothalamus. We are all synaesthesics when we are born; it is the death of cells that takes place during the normal development of the brain that gives rise to sense islands. In the uterus, there is a spectacular growth of neurons and they all compete to establish synaptic connections with the others. Those that don't manage it die. Between the ages of one and two years, the neurons are pruned back. It is believed that in synaesthesics there is a failure in this pruning process, so the synaptic connections between the different sensorial areas remain intact throughout their lives. Synaesthesics hear a dog barking and see a series of brown and grey triangles that reproduce upward and at the same time fade away. It is a little like fireworks: they appear, they remain suspended for a second, and then they disappear.

It is difficult to understand why evolution has not made all of us synaesthesic. If we had conserved the situation of the first months of life, we would enjoy undeniable advantages. Some years ago, Richard Cytowic told me, in his house in Washington, D.C., about the case of a Canadian TV producer who did not use a datebook in order to remember all her appointments, schedules, the people going to her program, the recording team, et cetera. "How do you manage it?" Cytowic asked her. "I have in front of me," she answered, "a cork board with colors and shapes and I put each thing in a certain place, the way the postman puts the letters in the letterboxes, and in that way I remember." It's a great advantage of synaesthesia. Synaesthesics usually have prodigious memories because the colors, smells, and shapes make them remember things. A synaesthesic would ask me what my name was and say, "You're called Edward. As green is an E, I know you are Edward because you have a green name."

Many people ask themselves if they wouldn't go mad with so much information. It is as if a blind person were to say, "Wherever you look, you always see something. Don't you go mad, seeing so many things?" It's a question of textures of reality.

Richard Gregory came back into his workshop, beaming from ear to ear. He had found his impossible pieces of wood, and the question was obligatory:

E.P. Are there things impossible for the brain to solve? I mean, perceptions that the brain cannot figure out and so gives up?

R.Gr. Yes. Years ago, I invented an impossible object, made from three pieces of wood. When you look at it from a certain perspective, the brain presumes

Fig. 2. Impossible Object Perceived.

that the pieces at the top touch each other. The eye perceives them as being in optical contact, so the brain assumes that they are at the same distance. Though generally optical contact corresponds to real contact, in this case it is a false supposition. The brain constructs a hypothesis, if you like, of the object based on a false premise, hence the paradox.

E.P. So in a way, when we obtain images of objects, the brain draws up a hypothesis on what that object is and then it tests it out, contrasting it with the senses, right?

R.Gr. Yes, I think that perception is a hypothesis. The problem is that perception has to work very quickly, in a tenth of a second, so verification is not very exhaustive. If we touch an object and we realize that we were wrong, the visual brain still cannot see it properly, because the hypothesis the brain creates is only partially related to intellectual understanding. Part of the brain understands and the other part doesn't.

E.P. We can even see things that don't exist, right?

R.Gr. Of course. We create hypotheses starting from very little information, beyond the evidence, like in science. From a few observations, we construct a large model of the universe or a model of the objects around us. Sometimes, this generator of hypotheses starts from zero and creates a fantasy, as happens, for example, to schizophrenics, who generate mistaken hypotheses.

E.P. So, perceiving is not only seeing, but also experimenting, even with the brain hallucinating!

R.Gr. I think that perceptions are controlled hallucinations. There is sufficient information from the eye and the senses to control them, but it is a creative process of fantasy, of fiction, that is normally under the control of the sensorial signals.

E.P. Allow me to show you another impossible situation, in which the mind gives up trying to find a definitive solution. I'm referring to the famous Necker cube.

R.Gr. Yes, I would say that here there are two hypotheses that the brain creates: one is that the front face is the one that is unshaded in the figure on the left, and the other one is that represented in the figure on the right. And here, there is no evidence to say which one is correct. In both cases, your perceptive reality is the one that is active in your brain. In this case, only two possible perceptions exist so you can flit from one to the other. In any perception there are always other possibilities just waiting to be selected.

E.P. I have a question that I've been looking for an answer to for some years. My consolation prize is that Richard Feynman, the Nobel Prize winner, sought the answer. When we look at ourselves in the mirror, we see our image in reverse. If I pick up a pencil in my right hand—many people don't realize this—and I look at myself in the mirror, I see myself reflected with the pencil in my left hand. Is it true that after completing a self-portrait, Rembrandt realized that he had portrayed himself with the paintbrush in his left hand?

R.Gr. Yes, he had to paint the self-portrait again because he realized he appeared to be left-handed when in fact he wasn't. It's incredible. A mirror is completely symmetrical. The question, as you said, is why it is turned round horizontally and not vertically if there is no optical turn. Many theories have been put forward: one is that the asymmetry is due to the fact that our eyes are separated horizontally. It has also been suggested that we have a mental rotation: that we imagine—or see—ourselves in the mirror and then we mentally rotate it to see ourselves face to face. But these theories are wrong. The question is why the mirror turns the reflection round this way and not the other way.

E.P. And what is the answer?

R.Gr. I think the answer is as follows: if you work in optics, you will say that it is a problem of optics; if you are a psychologist, you will say that it is the brain that turns it round; since Plato, an endless number of absurd hypotheses have been put forward. The answer is quite simple. To look at an object in a mirror, we have to turn the object such that it is in front of us. So, the top right-hand side becomes the top left-hand side, simply due to an asymmetrical rotation of the object because of gravity. The mirror reflects the rotation you have imposed on the object or on yourself in order to be in front of the mirror.

E.P. It would seem that the brain has strategies to avoid constant stimulation. We get dressed. Then we are dressed all day long but we're almost unaware of it, right?

R.Gr. Yes, absolutely true. It's called adaptation; becoming accustomed to a situation begins in the peripheral nerves. As you said, once you put your clothes on you hardly notice them, because otherwise the brain would be constantly bombarded by unimportant information. When an important change occurs, the brain's hypotheses have to be updated. If we constantly hear the clock, after a while we stop hearing it, we don't even hear it chime the hours, because it's not important. We know it will happen, it is not an announcement of anything new that must dealt with.

E.P. We're talking about science and not magic, but what relation has existed between science and magic in the course of time?

R.Gr. I think that science was born with magic, to the extent that it is a psychological explanation of the physical world. To explain the storms, it was said that the fury of the gods was unleashed. A psychological explanation was given for a physical phenomenon. But I think it is very worrying that right now, as Harry Potter demonstrates, children love magic, it fascinates them. A magic wand is more attractive to them than a telescope, a microscope, or a spectroscope. I think it is very sad, because science is absolutely wonderful. I wouldn't use the word *magic*, because it works; science presents marvelous questions and incredible answers, and explains the universe much better than magic..

E.P. One last question. You are practically the inventor of a profound reflection on perception and images of objects. After so many years observing, investigating, thinking, and writing about this, do you think that all perceptions are illusory? Or, with the years, do you feel that the outside world, the universe, is more or less what the brain generates?

R.Gr. A very interesting question. I think yes, there really is a universe out there. It is absurd to say everything is a dream. If everything were a dream we would lose the word *dream*. Words have to have some contrast in reality to have meaning. If you say that everything is an illusion, the word *illusion* disappears, it loses meaning. The question is, what is the truth of reference, based on which we say that something is true or is an illusion? I think that perception uses a commonsense physics, a "kitchen physics." In quantum mechanics, by contrast, or in the theory of relativity, physics describes a completely different reality. The reality of perception is a commonsense physics—the physics of touching, picking up objects, like foods. Perception must be applied to the world in order to cope with dangers, food, our companions—they are the only reality for perception.

E.P. The brain does not exist in order to find the truth, but rather to survive, you say.

R.Gr. Yes. Societies need an agreement in order to work collectively on the construction of a house or the manufacture of kitchen utensils. We need preconceived ideas in order to survive. Social beliefs enable us to carry out coordinated, agreed upon action. None of this has any relation to an absolute truth.

— 16 —

Dreaming to Learn

Interview with Nicholas Humphrey

The role of dreams is to throw us into
extraordinary social situations.
—Nicholas Humphrey

In his book *An Experiment with Time*, published in 1927, the British aeronautical engineer William Dunne details how the imagination mixes the future and the past, moves back and forth, in the same dream. Like Paul Davies's time machine, dreams break the barriers of space and time.

For years, there was no way of scientifically proving which of the two major currents of thought on dreams was correct. Science traditionally has argued that the brain, constantly bombarded by sensations, logically diverted

---

Nicholas Humphrey is the School Professor at the Centre for Philosophy of Natural and Social Sciences, London School of Economics.

all sterile information into the subconscious, from where dreams were generated. The brain's main objective is to distinguish the information necessary for survival from what is irrelevant. To scientific eyes, therefore, dreams are no more than an accumulation of waste DNA that remains deactivated in the course of evolution. Sigmund Freud (1856–1939) and Carl Jung (1875–1961), the two giants of psychology, laid the bases of the opposite line of thought. Freud suggested that dreams were an interpretable manifestation of the individual subconscious, whereas Jung emphasized the collective unconscious. Today we would call it the planetary brain. Meanwhile, science is making only slow progress in the analysis of dreams.

The application of modern magnetic resonance techniques has blurred the supposed clear separation between the world when we are are awake and that of dreams, the starting point for both traditional currents of psychology. The first truly intriguing finding that gave dreams a certain profile of reality and reality a certain profile of dreams was the realization that the same group of neurons is activated when we perceive a reality and when we imagine it. The brain doesn't seem to distinguish the visual images of dreams from the perception of reality when we are awake.

The second warning sign comes from experiments by neuroscientists on "lucid dreams," which suggest that the time elapsed in dreams is real time. Dreams are not condensed. They take place on measurable time. Too many similarities exist between dreams and the world of perception to dismiss them as a "junk DNA." Perhaps science is corroborating the Hindu conception that grouped three phases: waking, sleep without dreams, and dreams. They are seen as the same process together in consciousness.

At the same time, science fills the gap with regard to the function of dreams. It would be paradoxical or unacceptable for human beings to dedicate three full years of their lives to dreaming without any purpose. As Nicholas Humphrey, one of the most prolific and knowledgeable psychologists of our age, explains, "The Darwinian theory of evolution would never permit such waste." Research into dreams has only just begun, neuroscience almost unanimously accepts relevance to evolutionary learning and the development of the ability to solve problems, a central idea in Nicholas Humphrey's thought.

Nicholas Humphrey studied the behavior of gorillas in Rwanda with Diane Fossey and was the first one to demonstrate the existence of blind vision in monkeys that had suffered brain damage: the ability to respond to visual stimuli despite not being aware of having them.

Eduardo Punset: It seems that in the course of human life we dedicate an average of three years to dreaming, not just sleeping. Is this a waste of time or does dreaming fulfill some function?

Nicholas Humphrey: It would be very strange if all that time served no purpose. I think that dreaming performs a very important psychological and biological function as a part of our evolutionary heritage. Almost certainly, it is more important in humans than in other animals. Other animals may have small dreams, but I doubt they are those extraordinary narrative dreams that are like stories.

E.P. So, when my dog goes to sleep on the sofa and starts barking, she is not really dreaming stories?

N.H. Perhaps she feels an emotion related to hunting or chasing a cat, but nothing like human dreams, not even those of a six-year-old child. Both children and adults, when they dream, are heroes of their dreams, which are much like novels full of extraordinary events. Our dream adventures are crucial, though many scientists disdain the narrative content and believe that dreams can be explained in terms of neurochemistry, artificial intelligence, and computing. Dreams are as important in human lives as the reading of a novel.

E.P. Today's humans, *Homo sapiens sapiens*, are direct descendants of Cro-Magnon man. I have always thought that what differentiates Cro-Magnon man from its immediate Neanderthal ancestors is not so much language as fantasy and narration that constantly emerges in dreams. The need to create a fable around important issues is truly surprising—as if we were seduced only by stories and tales.

N.H. Yes, in that regard dreams are paradigmatic stories that draw on the first stories we hear. We get used to tales by means of the stories we construct in our dreams. Our first dreams are simple: many heroes in children's dreams are other animals, because people seem too complicated, whereas rabbits, cats, dogs, and tigers have simple emotions. So children begin to formulate ideas about what it means to be an active organism, have emotions, feel fear, hope, relief, or pain. I think that the main function of dreams is, precisely, to test out feelings, to put them into extraordinary social situations in order to learn how our mind works, how we react to strange situations that perhaps we have never experienced, but which might happen. I give you a magnificent example: Midwives often dream of giving birth, even if they never have had a baby. Midwives, women who perhaps may not be mothers, every day deal with women who experience the extraordinary feelings of maternity. I have interviewed midwives who say their dreams are crucial, because they give them an understanding, a perception that otherwise they would not have had.

E.P. Nick, some neuroscientists told me that they have observed that the neurons of premature children are activated when they dream, and they spend almost eighty percent of their time dreaming.

N.H. I think that is an exaggeration. The physiological signs such as the rapid eye movement (REM) when we dream, also present in other mammals, do not mean that the dream is a true experience. REM doesn't happen until the child is a little older. If we wake a two-year-old when he shows those brain waves and we ask him what he was dreaming or thinking about, he will answer simple things, that he was frightened or was very happy because he was eating an ice cream. By the time the child is four years old, his dream becomes a narrative, a story that can be dramatic and wake him up. "I lost my mother, I didn't know which way to run, I found her but inside a cave and I was looking for my father, but he'd fallen off a cliff and I was very frightened. I ran and ran and in the end I woke up." This kind of story, often much more elaborate, teaches the child about loss, danger, and anxiety. As the child grows, dreams become increasingly elaborate. He or she experiences falling in love, losing the loved one, betrayal. Through dreams, we learn about real life even if just in the form of a play. I don't use the word play figuratively, but rather like the theater. From *Hamlet* we learn about betrayal, incest, hate, and revenge. Plays are narratives that resemble those played out inside us night after night in the form of dreams.

E.P. Fortunately, this process of self-learning is different from what Aldous Huxley imagined in *Brave New World*, when he wrote that every night our descendants would have to see a recording to know what they had to do the next day. Or is it similar?

N.H. I think that, in part, you are right. We can't pervert the mind of a man or woman so that it works that way. The more we learn about the possibilities of human beings, the richer and more successful we are as what I call "natural psychologists." Almost everything I have written in the past twenty-five years or so deals with this extraordinary ability to understand other people's "natural psychology." We human beings are magnificent psychologists.

E.P. Because we can see what happens in other people's minds?

N.H. We read their minds. I concentrate on what you are thinking and I start to understand your motives, your experiences, your feelings, and your needs.

E.P. And not only that, then you can manipulate and influence me.

N.H. Yes, I can manipulate you but I can also help you. That is called the social function of intelligence. Years later, it was given a different name, which is not mine: Machiavellian intelligence. I think it's a great mistake, because the concept of Machiavellian intelligence suggests that the intelligence is used for manipulation, cruelty, and competition. I think that the social function of intelligence is very important because it makes love, compassion, and a positive social life possible in a way that is not possible for other animals. Some of my colleagues think as I do, that human beings are the only animals capable of this kind of perception. Not even chimpanzees or dogs can.

E.P. We'll return to this subject. If I've understood you, you used a very important word, *re-creation*, to refer to dreams, apparently because dreams simulate events or processes that you might encounter in real life.

N.H. Simulation is crucial because, of course, dreams are not something real and therefore do not involve dangers or the consequences of things that happen in real life. Exactly like children's games: while they are playing—

E.P. They develop skills—

N.H. Yes, they develop skills and they are in training. They develop the concepts that accompany types of relationships. When children play at being enemies, friends, doctors, nurses, or pirates, they learn crucial psychological concepts for their subsequent success in life.

E.P. We use dreams in the way a pilot uses a flight simulator to learn how to fly, in other words.

N.H. Exactly, that is a very appropriate analogy.

E.P. And what else? Because the time we spend dreaming in the course of our lives is the equivalent of three years.

N.H. Yes, we may dream for even more than three years. This demonstrates the importance of the function of dreaming. How many years do children play before they experience the real games of life? You should see my children, all they want to do is play. . . .

E.P. What is your opinion of "precognitive dreaming"? Dreams can be used to predict things that will happen. Do dreams have another function in addition to the one you suggest?

N.H. Probably it is inevitable that people seek and then find elements of their future in dreams. We humans are very superstitious. We look for all kinds of sources of information about the future. Prediction of the future would put us at an extraordinary advantage. Since the dawn of the history of mankind, in all ages, people have sought signs of the future both in the outside world and in dreams. Dreams are so extraordinary and so open to interpretation, as Freud and Jung studied. They contain symbols, colors, forms that lead to an infinite number of interpretations. By invention, we construct stories that fit in with the future. If you examine a dream, you can always believe that it contains a prediction of the future. I think it's an illusion and, like all illusions of precognition, it does not imply paranormal abilities. Though it's not surprising that we believe in those things because we want to believe them. We humans are fearful creatures, we need to know our future.

E.P. We are vulnerable in a very complex world.

N.H. Yes, the world frightens us and from the time we are born we try to make sense of it, predict it in order to understand who we are, where we are, and where we are heading. Dreams offer that possibility. But I think they are an illusion, like psychoanalysis, though very attractive because they correspond to our search for meaning.

E.P. Some years ago, you worked with a monkey called Helen who, like humans, had a clear preference for certain colors. She loved blue and hated red.

N.H. I know two stories about monkeys. Monkeys have very marked preferences for certain colors, but Helen was a special monkey. . . .

E.P. She was blind.

N.H. She wasn't blind. That is what was extraordinary. But she couldn't see colors. Normal monkeys have incredible feelings about colors, much stronger than those we believe human beings have. Monkeys cannot tolerate red light. If you place a monkey in a room with red light, it will shake and crouch down. If you expose it to blue or green light it will calm down.

E.P. But you know that if we ask a human about the same choice almost 90 percent of people prefer blue?

N.H. Yes, it's true. If we talk about very general preferences such as what kind of light would you like to have in your bedroom or under what kind of sky would you like to live. Another thing in pathological cases is that when someone suffers a certain kind of brain damage, he or she becomes much more sensitive to colors. A paralysis disorder exists that affects the cerebellum, located at the back of the brain. The people who suffer from it shake and move their arms. In a room with blue light or if they wear glasses with blue or green lenses, they calm down. With red glasses their agitation intensifies until they fall to the floor. It's purely an effect of the color.

E.P. Colors must be much more important than we think.

N.H. No doubt. They have been important throughout history. Some animals, like monkeys, demonstrate marked preferences. I think that one thing that happens with humans is that colors have begun to lose their meaning in the lives of civilized people. Colors no longer carry the biological implications they had in the real world. Bright colors—reds, yellows, blues—surround us in any shopping center or at any traffic light. They have lost the meaning they had for primitive* animals. For our ancestors, red was the color of fruit, of poisonous berries, or of the sunset. It meant danger because the most dangerous time of day was the night. One had to be on guard when the sun and the sky turned red. One had to be alert and watch what was behind you. This I discovered with the monkeys: they became very nervous and looked around them when the sky turned red. When the sky was blue, the predators were sleeping because it was midday and the monkeys could relax.

E.P. You imply that animals react to colors in the same way as our ancestors.

N.H. I think they are similar to what we were like a million or two million years ago.

E.P. What do you think of conscious dreams, when you dream but you are aware that you are dreaming, half asleep, "lucid dreaming?"

N.H. Yes, lucid dreaming. I have never had a lucid dream, so I only know about them from other people's experiences. All dreams are conscious: we explain our feelings when we dream as if we had experienced them. But "lucid dreaming" is an extraordinary type of dreaming in which he who is dreaming is aware that he or she is dreaming. He can control that dream and guide him or herself through it. So it has been suggested that if normal dreams can help us to learn and perform a role of enabling us to explore reality, lucid dreams are even better, because in a lucid dream you can build a narrative and place yourself in a specific story as if playing a computer game in which you control the events.

E.P. But unfortunately you can't turn the switch off.

N.H. No, though some people are convinced that we can be trained to have lucid dreams, with a device that measures the movements of the eyes and emits a slight electrical discharge to tell us we are dreaming. And with a bit of practice you become aware that you are dreaming and you have lucid dreams.

---

*The term *primitive* always needs explanation; it usually is a meaningless pejorative and should not be used. It may mean earlier appearance in the fossil record, or with simpler physiologically or fewer metabolic reactions, or less morphologically complex or fewer genes in the genome. We recommend replacement of terms like *primitive, lower, superior, higher*, with more meaningful descriptions.—LM

E.P. Do you think that we will ever obtain other benefits from dreams, in addition to rehearsing or training? Something more sophisticated, perhaps?

N.H. People often ask if the human mind can be improved, so it's a very interesting question. My intuition is that the human mind is the result of such a complex design, developed over the course of the past six million years of evolution—I'm referring to the moment we left the chimpanzees behind—that our mind is extraordinarily perfected, and it would be very difficult to improve. Anything that changes it or alters it might appear to be an improvement, but it may have dangerous consequences in other areas. I think it's best to leave it as it is, in its basic form, and be thankful.

E.P. Nick, I want to ask you something that fascinates me. There seem to be two phases in dreaming. In the first, we are capable of moving our body while we sleep, but we are incapable of remembering what we dreamt, or if we dreamt. In the second phase, the REM phase, we are capable of remembering what we dreamt, not capable of moving our bodies. We are paralyzed.

N.H. Well, not exactly. In the first phase, when we can move, there can't be much activity because we would fall out of bed. When we enter the deep-sleep phase, the REM phase, the body is paralyzed because otherwise it would be very dangerous. The lack of movement only affects the neck. We can speak and sometimes move our eyes or cry. But, except in very rare cases, we don't leave our beds to go downstairs or out the door.

E.P. Because we could get killed.

N.H. It has happened. A pathology that stops the appearance of this inhibition enables the body to move and act based on dream. People have been in these circumstances. They have felt impelled to commit absurd crimes, extraordinary acts. They are not responsible, in a way, because they don't intend to do it. In two or three legal cases in the United States a person has been declared not guilty of homicide because he was in a state of profound sleep when he did it!

E.P. Do you think it is good to dream?

N.H. The answer is yes. Dreaming is good, otherwise we would not spend so much time dreaming. Would you like to stop dreaming? Perhaps occasionally you would prefer not to remember your nightmares, but isn't it marvelous that we spend the night hours inventing incredible fantasies that transport us to places and into relations that otherwise we would never know?

E.P. In dreams our reasoning is more complete and sophisticated, you say, than when we are awake. I think that evolution should make us spend more time dreaming than thinking.

N.H. I think that dreams are a kind of very sophisticated consciousness, one of the greatest achievements of the human mind. When we are dreaming the mind is in one of its most creative moments, of that there can be no doubt.

E.P. If dreaming is one of the most sophisticated activities of the human mind, then dreams cannot be useless.

N.H. That's what I think, and I hope I've convinced you. If something is related to the human mind and the body in an intelligent way, it must correspond to some purpose. And, what's more, dreaming is a great source of pleasure. Darwin's natural selection doesn't lead to anything in vain.

— 17 —

Music and Talk

Interview with Diana Deutsch

Musical perception depends on the
speech acquired in childhood.
—Diana Deutsch

Perhaps an awareness of death is exclusive to humans. But, returning to language, and after listening to the song of a nightingale or the singing of the whales, the question arises whether music is a language. It is undeniable that music and language are similar. Music is a prolongation of language, or at least it is born from the limitations of the ability to speak. As language is a digital capacity of the brain, sooner or later hominids turned to music and art to express nuances that are difficult to express in the digital logic of zeros and ones.

Diana Deutsch is Professor of Psychology at the University of California, San Diego.

That the origin of music dates from a stroll by Pythagoras around the polis is hard to believe. More than two thousand years ago the Greek philosopher, walking along a street in Athens, heard a sound coming from a blacksmith's workshop that seduced him. He wondered why this sound fascinated him so much, and entered. Two thousand four hundred years later we think we know why a sound is pleasurable. Diana Deutsch made me question my certainty when I asked her about this in San Diego, California. Diana Deutsch is the world's leading authority on the physics and physiology of music. A graduate in psychology, philosophy, and physiology from Oxford University and a doctor of psychology from the University of California, she investigates the perception and memory of sounds, in particular, music. She has discovered a large number of musical illusions and paradoxes.

Diana Deutsch: Pythagoras came to the conclusion that pleasurable sounds are those whose components are related to string lengths that have a very simple numerical proportion, such as two to one. Since then, people have wondered why we like the so-called consonant, or harmonious, sounds more than dissonant ones, with more complex proportions. We still don't completely understand this. But we have discovered other interesting things about the perception of music.

Eduardo Punset: I have a question related to the famous Fibonacci series, which you know well: one, one, two, three, five, eight, et cetera, which led to the golden section. Here there is a proportion of zero to six that can be applied both to architecture and music. Beethoven and Béla Bartók apparently used the golden section. Why are those proportions so fascinating? Is it a miracle? A genetic matter?

Di.D. It's a mystery. We don't know. The reason behind the perception of music and all the arts is still a mystery. Many people believe they can judge aesthetics, but the reasons are, scientifically, unknown.

E.P. I read an essay by you that I thought was fantastic. Correct me if I'm wrong, but there seem to be acoustic illusions in the same way that there are optical illusions. The brain works on its own account. One classic illusion is the famous Necker cube: it is ambiguous, it can be interpreted as being in different positions. Have you demonstrated the existence of acoustic illusions?

Di.D. I was very surprised at the finding, because I was not expecting it at all. It is a very powerful illusion. Imagine a musical score at the top of this page: part of the space—which would correspond, with headphones, to an ear—receives a very disordered sequence, and the other part of the space too. But in the lower part of the page what is really heard would appear. It is completely different. It is as if the brain rejected the disordered models and reorganized the space and sound so that identical melodies are heard. When a channel is presented on the right of the listener, the one on the left does not hear the two channels that are being played, but rather conceptually reorganizes the sounds in such a way that it seems that all the high-pitched sounds come from one loudspeaker and the low-pitched ones from the other one. It is very strange. Right-handed people tend to hear the high-pitched sounds from the right loudspeaker and the low-pitched ones from the left loudspeaker, independently of the position the loudspeakers are in.

E.P. Right-handed people perceive the notes in a different way from left-handed people?

Di.D. Exactly. Right-handed people tend to hear high-pitched sounds on the right and low-pitched ones on the left, from wherever they come from. But left-handed people, as a group, tend not to hear in that way. Left-handed people vary greatly and, statistically, there are more models of perception among left-handed people than among right-handed ones—

E.P. Excuse me for interrupting you, but why? Are the neurons responsible for hearing the music different from the neurons that locate where the music is coming from?

Di.D. This is part of the problem. Left-handed and right-handed people have different forms in the organization of the brain. The vast majority of right-handed people tend to have speech represented in the left part of the brain, while left-handed people, as a group, are very different from each other. Therefore among them there are no consistent differences in the perception of the source of the sound which statistically correlate with their dexterity.

E.P. What happens in orchestras? Is the position of the musicians taken into account?

Di.D. Not only do right-handed people tend to hear high-pitched tones on the right and low-pitched ones on the left, but there is also another configuration. If you have a special configuration in which high-pitched sounds are on the right and low-pitched ones on the left, you can hear these models more clearly than when the low-pitched ones are on the right and the high-pitched ones on the left. The positioning of the musicians in an orchestra is interesting. From the musicians' point of view, the instruments with a higher-pitched register are located on the right and the lower-pitched ones on the left. In the violin section, the first violins are to the right of the second ones, which are to the right of the third ones, and the violoncellos are placed to the right of the double basses. In the wind instruments, the trumpet is to the right of the trombone, which is to the right of the tuba. This positioning is the result of many tests, of a process of trial and error, which leads to the most perfect possible positioning of the instruments, which is translated into the best possible performance, because, in the final analysis, the musicians must be able to hear each other in the best possible conditions to play the music in the best way possible.

E.P. This has been an evolutionary process?

Di.D. I think so. Orchestra conductors simply tried out different positions of the instruments, and little by little this positioning evolved.

E.P. And what about the audience? Is the ideal positioning for the orchestra also optimum for the spectators?

Di.D. That is the heart of the matter. From the point of view of the audience, the positioning from left to right is the reverse, like an image in a mirror. Therefore, from the audience's point of view, the instruments with the highest registers are placed on the left and the low-pitched instruments are on the right, so for the primarily right-handed audience it is the worst possible positioning for perception. There is no solution, because if we invert the positioning of the orchestra the musicians will not hear each other well.

E.P. A colleague of yours suggested that the solution was to retain the orchestra positioning and hang the audience from the ceiling!

Di.D. The theoretical solution is unacceptable to the listeners. Other suggestions: they should hang the orchestra from the ceiling or only allow in left-handed listeners. I think these are rather far-fetched ideas. No solution has been found; it's a paradox.

E.P. You wrote about a clash between the orchestra conductor Arthur Nikisch and the composer Tchaikovsky. Nikisch apparently did not like some parts of the musician's symphony. Your interpretation is that perhaps Nikisch didn't like them because he had a good ear, very precise, and at that time the musicians were not positioned on his right.

Di.D. Yes. I recount the legend that Tchaikovsky and Nikisch argued, shortly before the premiere, about the last movement of the sixth symphony. They clashed because the theme and the accompaniment alternated between the first and second violins, which at that time sat opposite each other on the stage. Nikisch wanted to restructure the piece so that the first violins played the theme and the second violins the accompaniment. Tchaikovsky wanted to keep the piece as he had composed it. The work was premiered as Tchaikovsky wanted.

E.P. At that time they probably didn't know that there are acoustic illusions. Let's consider the "tritone paradox," related to musical language, to the perception of music. You suggest we don't all perceive music in the same way.

Di.D. Exactly. I usually demonstrate my theory by playing examples of the tritone paradox and asking if some very simple models composed of two tones go up or down. The question is: when you hear it does it rise or fall? I will play the model and I will pause to ask you how you hear it. . . . Do you think it goes up or down?

E.P. Down.

Di.D. I think it goes up. In reality, there is great disagreement about this model and whether it goes up or down. I'll continue. I have four models and in each case I will ask you the same thing.

E.P. I think it goes down.

Di.D. Well, I think it goes up.

E.P. You can't be wrong, because you're the leading expert on the physics of music. I must be wrong.

Di.D. Many Californians hear this model like you do, but I am from southern England. The majority of the people in the south of England agree with me. What about this one? It's different. Does it go up or down? I think it goes down.

E.P. For me, it goes up!

Di.D. It is as if we were hearing exactly the opposite. And the third one, up or down?

E.P. Up, right?

Di.D. Down. And the last one?

E.P. I think it goes up.

Di.D. It seems we are still hearing the opposite.

E.P. Why?

Di.D. The paradox demonstrates that the way in which you hear this model is related to the sounds of your language, of the language you learned in childhood. Therefore, the geographical area determines how we hear the models. The people from southern England, like me, do not agree with the people from California. Since I demonstrated it, other investigators have corroborated this fact in other contexts: the perceptions of the people of Greece and Texas, or Minnesota and Sweden, differ entirely. Other studies have demonstrated differences in tone perception in other geographical regions. Another test consists of someone who speaks into a microphone for five minutes and his speech is recorded. Then samples of different tones of that same voice are made. The values in which there is the greatest incidence of tones are studied. This range of tones is statistically related to the way in which we perceive the tritone paradox, apparently. It is a very marked relation. So, for example, if someone expresses his or her opinion about the tritone paradox, I can guess what his or her range of tones will be. If I listen to a speech I can say—if the speakers are women (because I'm not very good at guessing the range of tones of men's speech)—how each will respond to the tritone paradox.

E.P. Diana, please don't tell me that women have a different perception of music from men …

Di.D. No, I don't mean that. I am more accustomed to the tones of women's speech than that of men. But you asked a very interesting question. In my studies of major differences in the perception of music, I noted the illusion of the scale, the different perception of right-handed and left-handed people, or the tritone paradox. The perception differences depend on the mother tongue and the geographical origin of the person. I have found no difference whatsoever that depends on gender. Perhaps one will be discovered, but I have found no difference in perception between men and women.

E.P. A neuroscientist told me that the same group of neurons responsible for food and sex are also responsible for music, so everyone likes music.

Di.D. Our perception of music is very significant: if two people perceive it in completely different ways, their aesthetic preferences will be very different. But beyond preferences, I think that some component of the taste for music is related to certain models of rhythm that penetrate in a very basic way.

E.P. Almost genetic.

Di.D. It's possible. I think that the majority of differences in musical perception are the result of the music we heard in childhood. The so-called critical period for the development of certain speech characteristics occurs between the first and second year of life, and even in the third. I think that the music we listen to, or to which we are exposed in the first years of life, greatly influences our musical perception as adults.

E.P. Diana, do you suggest that if children are exposed to good music when they are very young, when they grow up they like music, because it is almost like a genetic capacity, like that which leads us to eat, speak, or make love? If a person lacks the ability to perceive music, this is probably related to his or her early education or genes, right?

Di.D. Yes. But I don't rule out the influence of other factors. Those two factors, genes and early music experience, are very influential in musical perception. I would almost say decisive.

E.P. As a great expert in the physics and the psychology of music, why do you think that Western music, so metrical and rational, is so different from non-Western music, in which the melody and improvisation play a more important role?

Di.D. I am in favor not only of exposure to music in childhood but also to other linguistic models. Each way of speaking has special characteristics that are then recognized in music. This idea is very old: to appreciate music properly it has to resemble your language. This is very important because there is a great connection between the melodic flow of the speech we perceive and the music we like.

E.P. And what music do you like, Diana?

Di.D. Me? I really like Schubert, Bach, and the classical guitar of Andrés Segovia; I'm a great admirer of the music he played and composed. And Mussorgsky. Almost no one realized, but he made a great effort to get his music to imitate the sound of speech.

E.P. I think that the composer Mussorgsky said that the function of his art was the reproduction in the form of musical sounds not only of emotions but also of speech.

Di.D. Exactly. He said, "When I listen to any language, my mind—my brain—immediately starts to search for a musical expression for that speech." That was his aim and I think he achieved it wonderfully. If you listen to his songs knowing their aim, suddenly you connect to a different way of hearing and it really does sound like Russian speech; if you listen to it like that, you can recognize it. Though I don't just like his music because of that. He was an extraordinary musician with an incredible imagination.

Prehistory and Immortality of the Body-Mind

INTRODUCTION

We now know the secret code of DNA, but the discovery of the sequence of base pairs in the genetic material is far more like the floor plan than it is like the palace itself. Whether we truly will understand any life form is questionable. But to me it is clear that DNA is not enough, we must go far beyond equating life with huge linear strings of DNA sequences for 25,000 genes that code for a similar number of proteins, the molecules that do our body's work. Our brains, with their spanking new cortical folds, are lower Phanerozoic innovations, but our oxygen-breathing ability, which we inherit from our mother's egg, is far older. Our mitochondria lived free as bacteria, they took in oxygen from the air in the Proterozoic eon long before they teamed up with the rest of our cells. They did their oxygen-breathing respiration trick long before any animal or plant existed on the planet. And somehow the mitochondria in animal tissue cells are still involved in aging and death which, until now have been inevitable for all individual mammals. A mammalian body has thousands of millions of them, many per cell. You can believe whatever you wish but it is doubtful to me that we humans are just a little lower than the angels.

— 18 —

The Secret Code

Interview with Sydney Brenner

Children, all children, are naturalists.
—Sydney Brenner

Sydney Brenner is one of those rare originals who has drastically changed fields, lived on different continents, and followed his interests and passions with unerring good judgment from his, not his employers', point of view. With him, this approach has worked.

Eduardo Punset: There is life because there is reproduction of similar organisms, and they replicate similar life because of their genes. Does that mean

---

Sydney Brenner, Nobel Laureate in Medicine (2002), is a Distinguished Professor at the Salk Institute.

that replication is the signature of life? Is there a world without replication but just the sheer maintenance of whatever lives?

Sydney Brenner: I think there is more to it. If you look at genes, they are a description of the final organism. Replication occurs because it is hard to make complex organisms like animals by just copying them. Rather it is better to copy the description of the organism and pass that on. Copying itself is not enough for the life forms we can see. Encoding of the entire organism in a molecular description (that we call DNA) allows indirect propagation. When von Neumann—

E.P. The physicist?

S.B. The physicist, whom we might discuss later, dealt with the logic of reproduction and wrote the theory of self-reproducing machines. He very clearly separated the various components. First, the machine that did the work—the "effecter." In order to copy the effecter, an encoded description of it was necessary, "a program." Then a second machine was built by use of the program's instructions.

E.P. And then?

S.B. And then you copied the program. Then you took this program and inserted it into the machine. At that point you had two machines; each of them could do the same again. Of course, if the program was not accurately copied, through some inaccuracy, you got mutations. You could then have machines that led to more efficiency than the other machines. At that point you would get properties of evolution. Was that machine living? Or was it just a simulacrum? I think that it would be living in the sense of this definition of life. Perhaps something living must have added properties. When we think of a machine we think of the designer of the machine. Someone designed the machine, coded it on a tape, and let it go. The wonderful feature about living systems is that they evolved without a designer. This is still a great problem to be solved in the future.

E.P. This von Neumann model you mention really fascinates me. When I read about it in your book *My Life in Science*, I remember thinking it was the first time I had seen it written so that our audience could understand what the genetic code is. Genes are transmitters of information, yet you say something remarkable, mysterious: information is very different from matter. What do you mean?

S.B. Historically, when people started active science, they separated the living from the inanimate world. Effectively they thought there were two kinds of chemistry: the inorganic of the nonliving world, and the completely different kind, organic chemistry, the chemistry of life. When [Friedrich] Wöhler synthesized urea, typically a product of life, he exploded that idea. Chemistry was chemistry, from life or not.

E.P. And could be done in a laboratory?

S.B. Yes. Living things mocked the physicists and their quantitative analyses. Physicists always thought they had the "deepest" science, theoretically deep. They could not explain the phenomenon of life. Even until the beginning of the twentieth century many scientists, including Neils Bohr and Max Delbruck, thought new physical principles of life would be discovered. What challenged them was the growing awareness that physical matter and the atoms and molecules of life were made of the same chemical elements in the same kind of combination. The trouble with physicists is that they know no chemistry. Even though chemistry is a physical science, most physicists don't know any. We could call chemistry "very low-energy physics" as opposed to high-energy physics. The very lowest energy chemistry occurs in life. Biology is a selected part of chemistry. What was worried about has slowly eroded: life is not fundamentally different than chemistry. "The flow of matter" in biology led them to wonder if the matter of life could be the same as that of chemistry. Of course now we understand it is exactly the same. When bacteria make catalysts, here catalysts are enzymes, the chemistry is the same as in a chemistry lab.

Biologists became preoccupied with energy. They worried whether biological systems obeyed the second law of thermodynamics. If not, how did life escape from the laws of thermodynamics? The first law, the law of conservation, says matter is changed but not destroyed. Energy is transferred but not lost. The idea of entropy, that everything increases in disorder, seemed to contradict the orderliness of life.* Living beings seemed to increase in order. We now know about how energy is handled inside living beings. Unlike direct transformation of heat that irreversibly flows "into the cool" as in a steam engine, life's energy is packed into chemical bonds. Life seems to spontaneously become more complex, retain that complexity, and with mammals, even heat up and stay warm with no obvious "plugging in" or other energy flow. Heat cannot be stopped from spontaneous movement from hot to cold. Never, without expenditure of energy (usually a great

*For a more accurate explanation of the orderliness of life see Sagan, chapter 27. Energy spread, not "increase in disorder," is the appropriate qualitative language for description of entropy in the second law.—LM

deal of energy) can cold matter heat up spontaneously. Chemical bonds are broken and the energy transferred—in mammals energy comes from food. Biochemistry, the chemistry of living matter, worked for years on the problem of the source of materials to build organisms. Biochemists solved it. Then biochemistry faced how energy puts materials together. But the news is molecular biology, the concept of information, which is hard to understand. What is information in biology? Information in life is a branch of chemistry: the chemistry of genes. Information is embodied in the genes, not, say, in a computer, or little electrical charges, or magnetic fields. The genes are ordered long-chain chemicals. Biological information is embodied in a sequence of base pairs in the linear order in these chemicals. Another chemical problem, for now and the future, might be called the chemistry of organization. How are large complex collections of molecules organized to function as a single entity?

E.P. Were you not one of the first, if not the first, who said, years ago, "Let's not worry so much about where the energy comes from, and let's concentrate on the instructions, on the genetic code?" That fascinates me.

S.B. Yes. The "establishment" at the time, its conservative scientists, didn't think that "information" was a topic to be studied by chemistry. They had vague ideas. They worried about the sources of the components, the building blocks, and sources of energy to put them together.

E.P. When you first arrived in the United States in 1954 at Cold Spring Harbor you said, "Now I'm really getting into modern science." Have you ever wondered why the States was the place for modern science in 1954, and is still so today?

S.B. I was brought up in South Africa, where there was hardly any science, modern or ancient. Like many people who come from outside the center, I wanted to go to the center, which at the time was England, and especially Oxford. In Oxford, I tried to study that which had not yet been invented. I called it "self-theology." But I studied physical chemistry because I realized that the problems of biology at the end of the day would be chemistry. I then went to the United States because it was the growing country in science with a big scientific population, money, laboratories, and it still is. Probably in the subject of molecular biology the United States does 80 percent of the science in the world.

E.P. People who research Nobel Prize winners, of which you are one, say that among Nobel laureates, or even in science, generally there are two types of scientists. One is the visionary who looks into many things, and the other the "driller" who delves deeper and deeper into a single subject. But I read about your work, and find it very difficult to categorize you. What do you think?

S.B. Isaiah Berlin wrote a wonderful essay called "The Hedgehog and the Fox," which comes from a more or less similar Greek proverb. The hedgehog knows a lot about one thing whereas the fox knows a little about everything. They ask me if I am a hedgehog or a fox. I've decided I'm neither. Rather, I am an octopus. I want to know everything about everything, so as an octopus with eight arms I embrace many things.

E.P. Even strange ones, like when you looked at cells, you wondered about cell polarity—why they moved in one and not the other direction. Why do they face the wall from one side and not the other? You were intrigued by this years ago. Did you find the answer?

S.B. Not personally, but others have found it. We knew the first principle; that polarity is very clearly important. One end is different from the other. In many cases of development the answer is known. Archimedes said, "Give me a fulcrum and I will move the world." And I say, "Give me one polarity and I will explain everything, because polarity can be propagated."

E.P. After years of preoccupation with the genetic code, working with the nematode worm, *Caenorhabditus elegans*, the sequencing of genes, and then participation in the Human Genome Project, the time came when you decided to investigate complex nervous systems. That happened to Francis Crick, as well. The most complex system seems to be the human nervous system. Do you think any more is known about the brain and the rest of the nervous system than ten years ago? Than before that? How does it work closed inside, without light, and just receiving codes?

S.B. The brain is the most complex biological system that evolves, and the human brain is the most complex of all. The only brain that might rival that of humans, at least in complexity, is the brain of the octopus. For an invertebrate animal the otopus brain is extraordinarily complex. Octopuses are interesting to me; I plan to work on octopuses now to see a second, relatively independent, evolution of a brain.

E.P. In another branch?

S.B. Yes, on another branch of the phylogenetic tree. Octopuses are probably as clever as mice, but not as clever as humans. To return to your question: I've been interested in the nervous system forever, but for many years I just thought it too intractable. So when we decided that once the major principles of molecular biology were complete, both Francis Crick and I wished to pursue the nervous system, but we went in very different ways. He went straight to the human brain.

E.P. And consciousness.

S.B. Yes. He spent his life worrying about problems like "consciousness." I decided on a much simpler approach. I started with *C. elegans*. Its brain has only three hundred cells. It still took us a long time to make a "wiring diagram," to show all the nerve cells' connections. That is a very interesting challenge. It follows, in fact, another of von Neumann's prescriptions. To explain complex things, separate their critical attributes. The following question had not been solved for *C. elegans*, but might now be solved: can I separate the behavior from the wiring diagram? If I can separate behavior from the wiring diagram, I would consider the problem of that brain solved. Behavior is computable, I would argue, from wiring diagrams, which it should be. There is nothing else that can explain the behavior: the wiring, the pattern, the chemistry of the nervous system, and its responses.

E.P. You always said that in education or research, one reads diagrams to appreciate the different size scales. A microbe is not an elephant. I see that you have continued with that.

S.B. Yes. The key to understanding, even in the *Caenorhabditus* case, is that many different, simultaneous networks exist. We really need to come to grips with their fundamental properties. A student of mine simulated worm movement. He wrote a computer program that shows little worms crawling on the screen. I looked at his program, full of mathematical functions that generate sine waves, in the serpentine way his worms move. I told him his was not a simulation but rather an imitation. He had to be careful. So what would be a real simulation? I wanted to know about firing neurons that stimulate muscles. I needed the details, otherwise I could not study a generative function. I could not say that the wiring diagram when plugged with the right chemistry and interactions would generate the behavior. This challenging problem has not been solved. Most would say that this worm is too simple and therefore they are not interested. They are interested in complex behaviors: falling in love, regaining consciousness, or something like that. I believe if we can understand the wiring diagram relative to worm behavior we'll solve many more complex problems.

E.P. Other behaviors?

S.B. Yes. Other simulations, proper simulations. Nobody designed the nervous system, it does not resemble a computer program written by a programmer. The *C. elegans* nervous system evolved by accretion. This fact, that it is a product of evolution, becomes necessary to understand how it works.

E.P. We did one TV program on Redes called *Fish: Our Ancestors*. You also worked with the Japanese puffer fish and the fugu that belongs to the globefish family. Correct me if I'm wrong, but that ancestral fish has more or less the same number of genes that we humans have, but with a big difference, a difference in that one of our genes has a length of sixty thousand bases, and in the fish only five or six thousand. So if you could get a particular gene of that fish and find out its use, you might reduce time spent on the human genome sequence. So you did. Is this the future of evolutionary genetics?

S.B. In part.

E.P. Please explain.

S.B. One can choose any fish. That one was chosen because of its size. They were going to sequence the human genome, but I was correct: the technology was still unsuitable for such a large problem. Ultimately there had to be an improvement in technology. Ultimately the sequence was done in "factories" with hundreds of machines. I call the fugu fish genome the "discount genome." All this was borne out. Because there's less fugu fish DNA, it's much easier to study. As we predicted, the fish sequence proved to have great use in interpretation of the human genome. At the time nobody of course supported this. I'm afraid that is intrinsic in the kind of organization that I call Stalinist, where everyone decides how science should be done and who should do it, what they do and do not allow, the no-deviation policy where they will not support alternatives. Anyway, we got it supported, fortunately, and the fish sequence was done.

E.P. Are we nearer now?

S.B. Yes, nearer to interpretation of the genomes, but I'm afraid that most scientists, and especially computer experts, believe that computers will discover answers. I don't think so. I think we will need much deeper analysis of genomes than most people do today. The fish genome work has been very useful. It is a tool. The most challenging intellectual problem is whether or not we can reconstruct the past from the sequences. Genomes are related to recorded history, but we have to learn how to read them.

E.P. To understand how we evolved from fish to human?

S.B. Precisely. When in your mother's womb you were a little fish once. You had gills and everything. There is a very famous saying by Ernest Haeckel in the nineteenth century: "Ontogeny recapitulates phylogeny." In embryonic development, you went from a little fish to a little frog before you became human.

E.P. Thanks to the fish?

S.B. The fish started it.

E.P. What is in the future for this genome work? What will we be able to record about our past? Not only the past, but will it help to control our future a little better?

S.B. No. This is one thing that I realized when I traveled around last year at the fiftieth-anniversary celebration of DNA. I gave many speeches and listened to many talks about the "genetic improvement of man."

E.P. Germline genetics?

S.B. "Germline genetics." Improvement of heredity. I thought, "You know, this is not the way to do it." Man gave up biological evolution; today the technology is obsolete. If you look at what we accomplished in the last ten thousand years it has not been accomplished by genes. Rather, the accomplishments are by the brain. We have shifted: biological evolution produced the brain, but now the brain is moving to a different stage—toward cultural evolution. So even though the comedian Woody Allen says the brain is the second most important organ in the human body, he is wrong; it is the most important organ. I think improvement will be much better by cultural, not genetic, means as a product of our brains.

E.P. By technology?

S.B. By interaction with other people. By application to changing worlds. Rather than change the genome, we must change the world. It is one essential thing that if we were to study human sciences we need to see the continuity of biology, history, archaeology, and anthropology. I think that is the coming way of integration of our views of the world.

E.P. Do you think that from a genetic point of view, we probably are no more intelligent than sixty thousand years ago? What will model the future is a matter of technology, culture and the brain, rather than the genes themselves?

S.B. Yes. I also think that one tremendous feature of humans is very important: diversity. Everybody is different. The fact that we all look different was an important thing in selecting diversity. So I feel very strongly that we should do nothing that tampers with that.

E.P. Empower?

S.B. Yes, empower that diversity. Purification of the genome would be the wrong way to go because we would probably purify pretty bad things in our genomes.

E.P. Serious research done maybe in the early 1980s on homicide rates in Chicago reveal about nine hundred homicides per million people per year, compared to England and Wales, which was something like thirty. The conclusion drawn was that although the rates were different, the genders and ages of the people when the crimes were committed were identical. The "bad" genes are present. In calmer environments, the rates decrease; the worst environments facilitate violence. In other words, we get an easy formula that says genes play one part and the environment plays another part. I don't want you to, and I know that you won't come into the nature versus nurture debate, but knowing so much about genes, how do you see things from this point of view? What is the implication for the role of genes?

S.B. I think it is correct to say that genes build a baby; they do. We also know that mental illness has a genetic basis. But we also know that environment plays a tremendous part in humans. There can't be single genes for "homosexuality" or "criminal behavior." The situation is too complex. People like to have genes that will extend in a one-to-one ratio for behavior. I always give this example. A colleague of mine told me he had discovered the gene for obesity, the one that makes people fat. So my answer? "No, I discovered that gene a long time ago." He asked, "What gene was that?" I said, "The gene that opens the mouth." Of course to say that there is a gene to open the mouth is ludicrous. Almost the same as saying there's a gene for homosexuality or obesity. These are all complex developmental processes. You can kill them or make them defective one gene at a time, but there are many causes of any defect. To have a live being all the genes must function.

E.P. Many members of our audience are very young. In your work I find "clues" for people interested in research or studying science. I will mention two or three to you. You say that in your childhood you didn't realize that it was useless to ask people what you wanted to know, rather the best approach was doing. Educational reformers today talk about "learning by doing." Do you still feel the best way to learn is by doing?

S.B. One of the things you find today is that no one feels they can do anything unless they have a course of study that "covers" it. The assumption is that people who attend a course and pass the examination are qualified. The best way to learn a new subject (I still do it because courses are useless) is to get a book and start reading about the subject. Reading through it I then assimilate the knowledge there in my own way. Of course, "doing" is part of science; "doing experiments" is a craft that I find enjoyable in its own right.

E.P. Another clue you give people is you suggest someone from outside look at the problems in which you are immersed. Sometimes you forget and ignorance is needed.

S.B. It's all connected. If you know too much about a subject you'll say it won't work. So someone who is ignorant will try things. They will discover new things because they try them. The whole question of outsiders in a subject has begun to fascinate me because molecular biology was created by outsiders. It did not come from standard biochemistry but from physics and other sciences. A new science was created. A young person is always an outsider. You are still ignorant when you are young. For the young in science the best thing is to do it. I think every child, four or five years old, is interested in nature. The other day a lecturer said that when he was young he learned science because he took apart radios and learned to reassemble them. We did the same thing: we would grab a fly and pull its legs off. But we couldn't reassemble it! Children, all children, are naturalists; they want to explore nature. They go to school and schools kill their curiosity or their desires to act on their curiosity. Children at the age of five should follow their interests—which means they will forget about school and university, start Ph.D. study, and stick with what interests them for the next twenty years.

E.P. I was Minister for European Relations for a while during Spain's transition. I remember that whenever I said "Let's try this," I'd be told, "This was already tried, and no, it can't be done." I concluded from my experience that ministers should not hold office for more than six months!

S.B. Precisely.

E.P. They start to "know too much" and are useless. Another clue, you say that it is very advantageous to change jobs. You almost don't talk about it, but you refer to the misery of a fixed, permanent job. Is it not good for the individual or for science?

S.B. I say a fixed job and I mean that you stay in the same subject and you grow with the subject. Most scientists really become managers rather than scientists, at least in my field. There are still a few who don't.

E.P. You once said, "Monsters upstairs and idiots downstairs."

S.B. In the future people will have many jobs that they will start when young by doing research. Then they may change and study something else. Some people, I think, when they reach fifty will actually pay us to do science in the laboratory as a hobby, as recreation. I think to me it might be more interesting rather than going out to play golf or doing any of the other things people do, those who retire from important jobs.

E.P. Keep asking nature questions?

S.B. Yes, keep asking questions! I think one should go and do science for recreation or maybe become an undergraduate in French literature at age eighty!

— 19 —

Beyond the Human Genome

Interview with William Haseltine

We know now that practically all human health problems will be resolved one day.
—William Haseltine

I felt certain that the triumph of sequencing the human genome would bring immediate clarification of the role of genes in fashioning the body, understanding our relation with other primates, insight into disease, and other astonishments. But did it? Haseltine ought to know, and he does.

Eduardo Punset: Have you ever felt frustrated? When we succeeded in sequencing the genome we thought that we had discovered the secret of

William Haseltine is a professor at The Scripps Research Institute and President of Haseltine Associates Ltd.

life. Then you and others came along saying that although this was important we now needed to learn about proteins. They are much less static, they move around a lot more as they are more flexible, complex, and numerous.

William Haseltine: I feel both. The human genome is a marvelous achievement made by humanity; a piece of knowledge that now we all share. It is not only a current model, but it also may explain our past. We can analyze history in great detail, going back 3.5 to 4 billion years in history. Our DNA is a living fossil. This DNA, inside you and me, and which we now read and store in computers, is a text that has existed in a mutating and changing form for some 4 billion years. DNA is an immortal molecule: it has always been there without interruption. This is a marvelous reality.

E.P. It was already there 4 billion years ago?

W.H. The essence of life is the capacity to reproduce. When we see a beautiful countryside, a landscape covered in grass, trees, and flowers and the rocks beneath them, it may seem to us that the rock is permanent and the tree is perishable. This is a mistake. The rock will erode over time, but life will not be destroyed. Life has existed for more than 3.5 billion years, but rocks do not last this long. So where is the frustration? Frustration lies in possessing the keys and not being capable of applying them to solve problems pharmacologically. We know now that practically all human health problems will be resolved one day. Before we had this information we couldn't say that.

E.P. Then what exactly is human illness? Is it a fault in DNA? In protein-synthesis mechanism? Aging, or what?

W.H. Two or three basic categories of illness include infection. This occurs when the genes of another organism interact with our own and distort their functioning. It provokes the death of cells, tissues, and distortion of organ functions. The second category is intrinsic: something goes wrong in our own body, independently of any outside force. It often is something inherited. Third category is age. Today we redefine the notion of age: to a certain degree it constitutes a series of treatable illnesses. New medicine seeks to restore the normal health of a young person in the body through the application and knowledge of genes and proteins and other body chemicals. Expectations of this new medicine, instead of feeling frustrated, makes me realize that we are in a new world full of opportunities.

E.P. In the fight against disease, trauma, or aging, humans first used plants, then they extracted and made chemical substances, often from plants. Now we will use our own body chemistry.

W.H. It is a logical progression. Don't forget that not only plants are used, but also mechanical supports, such as wooden legs or false teeth. The new medicine will not only be confined to human parts, there will be a fusion with the material world, because of two main revolutions. We have already talked about one, the biological revolution, but at the same time there is a revolution in material science: the nanotechnological revolution. That is, where we used to use splints, now we will use nanotechnology. And is this in the long-term future? No, because even today there are people who have nano-implants here to enable them to see, or in their brains to help them move a cursor on a TV screen by using thought. The regenerative medicine of the future, as I see it, uses human components—our genes, our proteins, and our cells combined with nano-materials if necessary.

— 20 —

The Second Brain

Interview with Phillip Tobias and Ralph Holloway

I don't know any chimpanzee who has a sense of the future.
—Phillip Tobias

Mankind is perturbed, and that is why we were selected!
—Ralph Holloway

Phillip V. Tobias and Ralph Holloway maintain that the Neanderthals could speak, but it would be difficult to demonstrate scientifically. Three hundred thousand years ago there was no way of recording a meeting in a cave of a Neanderthal tribe. Nonetheless, it is important to answer the question of

---

Phillip Tobias is Professor Emeritus of Anatomy and Human Biology at the University of the Witwatersrand.

Ralph Holloway is Professor of Anthropology at Columbia University.

whether or not Neanderthals spoke, for several reasons: first, because Neanderthals lived alongside our immediate ancestors, and therefore are the nearest fossil men that evolution has provided. If we are unable to prove that they could speak, we conclude that our species was the first to use word language. This would distinguish us from all who came before. Our common ancestors, of humans *(Homo sapiens)* and chimpanzees *(Pan)*, are much more ancient than Neanderthals (also the *Homo* genus). They lived until six million years ago.

Secondly, if Neanderthals did not speak, then the genetic mutations necessary to make language possible took place in the Cro-Magnon variant of *Homo sapiens*, and not in the Neanderthal lineage. Alternatively, if these two species spoke they were not so different as to absolutely rule out their common descendants. The genetic capacity to develop speech would have been a Neanderthal characteristic also. Evolution does not seem to produce dramatic surprises in the short space of a few hundred thousand years.

The Neanderthal speech subject, still immersed in mystery, is one of the most fascinating unresolved questions about the evolution of mind. From the physiology of language, we will then consider the crucial question: the size of the brain and the content of the genetic expression necessary to be able to speak. An idea that at first sight appears far-fetched will seem plausible: two decisive moments led, in evolution, to human communication. The first step was taken by the reptile-mammal transition organisms two hundred million years ago. Forced to live in a relatively dark, shadowy world, they supplemented their extremely sophisticated daylight visual system with increased brain capacity that permitted the development of the senses of hearing and smell. The process that led to the largest brain ratio among mammals thus began in our primate ancestors.

The second transformation took over during the course of the past two centuries, the scientific explosion and the revolution in information technologies and telecommunications. The contemporary equivalent of the second brain of the mammal-like reptiles are computers. In mammal-like reptiles, the creation of the second brain never led to the disappearance of the reptilian brain. We still have it. Probably the same will happen with the incorporation of the external brain of computers and the Internet.

Did the Neanderthals speak? History was written by descendants of the *Homo sapiens sapiens* Cro-Magnons, namely, the victors. As the most widely recognized paleontologist, an expert on this period, Phillip V. Tobias, noted, history is always written by the victors: in this case, us.

Ralph Holloway, professor of paleo- and physical anthropology at Columbia University, New York, is an expert on the evolution of the brain and on primate behavior. He joined in the conversation in the lobby of the Hotel Colon [Columbus], which faces the cathedral at the edge of the Barrio Gótico, the gothic quarter of Barcelona.

Philip V. Tobias: I am certain that Neanderthals could speak. They had a highly developed culture, and required language to communicate it. The anatomy of their jaws differed a little from ours, but some *Homo sapiens* today also have jaws like theirs. The Neanderthals *(Homo sapiens neandertalensis)* were very cultured. It saddens me that today the word "Neanderthal" is so pejorative. What do you think, Ralph?

Ralph Holloway: I think too that the Neanderthals talked a lot. We don't know if they expressed themselves like you and I do. I'm surprised by their anatomy: the base of the skull, the supposed descent and length of the larynx. I don't like the subject very much. The question reminds me of a cartoon in which an unidentified hominid sculpts a rock with a chisel and a hammer. An enormous block. He chips bits off until he obtains an upright piano. Then he starts banging his head on the keys. He is a Neanderthal! I don't know if that is an exaggeration. Were the Neanderthals like that? We will never know if they were less intelligent than Cro-Magnons. We won't correctly answer that question until we travel in a time machine that perhaps the quantum scientists will build.

Eduardo Punset: A grandchild of mine, Miranda, was born a week ago.

P.T. Congratulations!

E.P. I tried to examine her larynx, this supposedly human characteristic, to see if it were true that a newborn's larynx is in a very high position so she can breathe and suck at the same time. A few months later, the larynx descends to a lower position so there's room for resonance. Humans supposedly are the only animals who choke. Is the complexity of the human vocal instrument a necessity, a requirement for language?

P.T. I say that we humans speak with our brains and not with our tongues. The difference is fundamental. We speak with our brains. The larynx, throat, lips, tongue—the organs of the anatomy of speech—are simply the executors carrying out the order from the brain. Therefore, we speak with our brains.

E.P. If it's big enough.

P.T. That is the other question that perhaps you would like to answer, Ralph. Was it the size of the brain that led to speech?

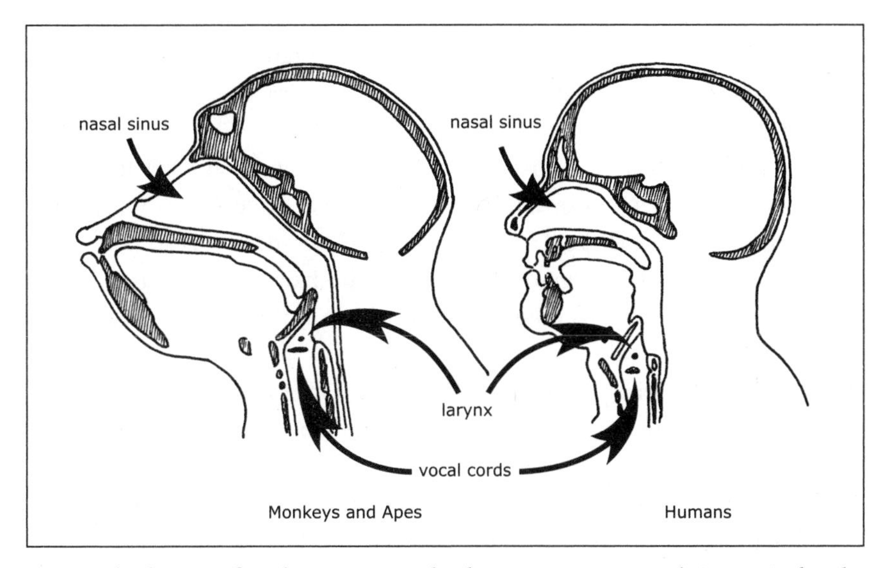

Fig. 3. The larynx of a chimpanzee and a human are compared, in particular the descent of the larynx in the hominids relative to the chimps and other great apes. This reposition helps preclude choking when speaking. This suggests that only humans can speak and not risk death from choking.

R.Ho. I doubt it. I think that to develop speech the organization of the brain cortex and some parts of the subcortex of the brain are more important. When I finished my doctorate, I studied a disorder, microcephaly. I found something that really impressed me. It demonstrates what I said. Microcephaly is a recessive disorder that leads to a very small head. The gene must be present in a double dose for this disorder to appear. The face and the rest of the body develop normally, but the brain stops growing early in the fetus. The person is born with a very small brain, sometimes smaller than that of monkeys. Microcephalics, some of them, are capable of speaking and being understood, although those who live are profoundly mentally retarded. Even so they are capable of speech. A theory in anthropology states that 650 cubic centimeters of brain mass are necessary for the development of language of human normalcy. The brain can be sufficiently organized to have the ability to speak even if it fails to reach an adequate size.

E.P. Perhaps the difference does not lie so much in speech as in writing.

P.T. For decades, people have discussed when speech arose, but no one I know of has raised the question of writing, Eduardo. Archaeological investigations show origins of writing and prewriting. Writing as such did not appear immediately, but rather graphic representations and paintings first appeared in caves in Spain, France, and South Africa.

E.P. When? Ten thousand years ago?

P.T. Yes, probably before. A work of art has been found that dates from sixty thousand years ago, on an archaeological site near Cape Town. Some paintings in caves in France date from twenty-five thousand years ago. The oldest is a recently discovered engraving that dates back over sixty thousand years.

R.Ho. In Australia, cave paintings from around sixty thousand years ago have been found.

P.T. That was, precisely, the start of writing. Little by little, like in hieroglyphics, the strokes of the paintings became more stylized.

E.P. Transforming into symbols?

P.T. Yes, symbols understandable by the community, which gave rise to writing, first in China, then the Near East and then in the so-called New World. In South Africa, a tribe called the Hottentots (the San Bushmen) engraved in a stylized manner, just like the Egyptian pictograms. My teacher, Professor Von Richtlow, used to say, "If the black populations had not reached South Africa, they would be a yellow race." And the whites wouldn't have left the caves. Those people, independently, rediscovered writing, or were about to.

E.P. It seems grotesque to see how fiercely we try to find the attribute that supposedly differentiates us from the other animals. The touted differences are proven wrong one after the other. First, it was use of tools, but then tools appeared almost entirely made by chimpanzees. The tool theory was abandoned. Then language, until it was discovered that dolphins also speak. Later the symbolic capacity was put forward. True, chimpanzees don't normally fly flags, but with regard to the symbolic capacity perhaps writing is the great difference between humans and other animals.

R.Ho. Language should be considered a cognitive activity of the brain that shares certain properties with other cognitive aspects. The appearance in fossil remains of stone tools manufactured in accordance with highly standardized models is extraordinary. These models cannot be justified by genetic mechanisms, the way the spiders can be classified into species depending on their webs. The stone tools demonstrate extraordinary patterns. Now, could they be manufactured without a high degree of social cohesion?

E.P. This degree of cohesion perhaps required the ability to communicate, right?

R.Ho. Yes, exactly. I mean, a high degree of social control was necessary, though not in the pejorative or authoritarian sense; rather, there was a considerable capacity to share experience. The brain at that time already demonstrates the symmetry of the two hemispheres, which favors the type of operation from left to right. It indicates the skill of the right hand. An archaeologist called Nick Toth examined the shards knocked off when stone tools were made. He reconstructed the manufacturing process of the tools and demonstrated that they were made with the right hand. Social behavior is the key that explains the evolution of these skills.

E.P. The differentiation of the brain between the two hemispheres, and therefore the location of language in one of them and the predominance of right-handedness in hominids, from when does it date? Two million years ago?

R.Ho. Yes, two million years. I read a summary of research by use of magnetic resonance techniques into brain models with Chinese and English speakers. They do four tests (I don't remember exactly) related to the perception of intonation, syntax, verb and noun structure. The Chinese brain areas lit up more than the English, because they pronounce more intonation in their language. The Chinese are not born with this brain model, but with the development of their language certain parts of the brain increase in size. The environment does affect the structure of the brain, and vice versa. All humans learn the language of their culture in the same period of time. A genetic component in language acquisition is irrefutable.

E.P. If human beings incorporate syntax into their language from when they are very young, can a chimpanzee order words?

P.T. There is a genetic basis for language. That linguistic skills are acquired shortly after birth is determined by the brain, which inherits certain abilities, sometimes very specific ones. In a family of musicians like Johann Sebastian Bach's, musicality may have been a specific factor of their genetic configuration, though of course what determined genetically and what results from environmental influence is very difficult to differentiate. Though the genes cannot differentiate the music of Beethoven from boogie-woogie, they can determine that one is more sensitive to music. Musicality, mathematical genius, or some types of serious illnesses abound in certain families. Certain individuals combine both genius and schizophrenia. Two recent Nobel Prize winners were schizophrenics. Einstein's son is schizophrenic. From an evolutionary perspective, could a gene for schizophrenia—which dates from one hundred and fifty thousand years ago or more—manifest indirectly in musicality, genius, leadership, and religiosity, more than in psychotic behavior? Religiosity played a very important role in the early evolution of culture.

R.Ho. Mankind was perturbed and that is why it was selected!

E.P. If we ask scientists who study brain evolution, like you do, you seem to not always agree on your answers. A colleague of yours, the physiologist from New York, Rodolfo Llinás, says that the brain appears only if it is necessary. For example, plants don't need a brain because it serves, in essence, to find direction and to anticipate certain events [see chapter 11].

P.T. I disagree with your pure theology statement or the old Lamarckian argument that because plants don't need a brain they don't have one. This idea is belief, not science, that "we mammals need a brain and therefore we have one." I'm afraid Ralph that I can't accept that argument.

E.P. Are we humans more intelligent than other animals because we acquired language, or did we acquire that ability because we are more intelligent?

R.Ho. I don't know. I don't think the question can be asked that way.

E.P. Did we acquire language because of our intelligence or because we already had language? Or perhaps, as you have suggested, organization and genes enable us to achieve language that, in turn, made us more intelligent than other animals?

R.Ho. Language contributed a great deal to improvement of social interaction and behavior. It's extraordinary, because it is based on an arbitrary code. That is, there doesn't have to be any relation whatsoever between the words and the reality, so the imagination can invent reality as it wishes and say what it wants, even if it is not credible. You can say, "A yellow hippopotamus looked at the blue-striped giraffe, but since he couldn't stand her, he kicked her tail." Imagination forms part of human intellect, but it leads us to sin. Invention creates great projects, like the religions.

E.P. Invention plays with the ambiguity that is characteristic of language. A popular saying in Spanish is "You get to understand people by talking to them," but experience demonstrates that language confuses us, too, because it is vague and interpretable. What do you think?

P.T. Many differences distinguish modern human from other animal communication. Yet it is indisputable that without language, especially spoken language—though sign and musical language probably appeared first—a sense of the future is inconceivable. Chimpanzees have no sense of the future beyond the immediate future. Satisfaction of its biological needs: hunger, sex, danger, tiredness, sleeping. Humans, in contrast, plan their futures, including very distant futures. We can forecast weather. Language makes possible forecasts and predictions. The subjunctive mode of language which does not exist in the communication of chimpanzees, gorillas, or orangutans is very important. As Chomsky said a number of times, we can't use language if we don't know grammar, because saying "I kill you" is very different from saying "you kill me." They are the same words, but the meaning is the opposite. And "You and me kill" implies a third meaning, but the words are the same.

E.P. It's like a genetic code, isn't it?

P.T. Yes. Combinations and permutations: "I," "you," "kill," or "kill?" are very different from each other, though the words are similar. Modern human language would not be the same without syntax and grammar. There are people who think that Noam Chomsky went too far in arguing that syntax and grammar are the essential bases of language, but without syntax and grammar human language would not be differentiated from that of the chimpanzees that shout, repeating sounds. Perhaps each sound means something different—"uh, uh, uh:" love, sex, or friendship; "hi, hi, hi:" anger, fear, or "a man with a rifle is approaching"—but they can't express anything in the future or in subjunctive. No more complex ideas. I think that the variations in the behavior of modern man appeared as a result of language.

E.P. Do we know what was the first word spoken by a primate ancestor of human?

R.Ho. Perhaps it whistled: "phew, phew!"

P.T. Or perhaps it just said "mama!"

R.Ho. Sex and food are the basic aspects of life.

E.P. And safety.

R.Ho. Only after sex, safety.

E.P. Do the sounds made by babies reflect primordial language?

R.Ho. It's an old argument of phylogenetic reproduction, but I don't believe it.

E.P. Let's not lose our thread. Professor Tobias, are we humans the only animals who imagine the future? I interviewed a scientist from the United States who researches how to displace the Earth's orbit in five hundred million years' time, when the planet will be too hot for life. Do you think that is an exaggeration on his part, even though the capacity to think of the future is a characteristic of the human condition?

R.Ho. Yes.

E.P. Really?

P.T. I'll add a caveat. We haven't yet fully investigated how whales, dolphins, and other large cetaceans speak and what they say. Scientists study the vibrations and the patterns. They think there may be surprises. The cognitive repertoire inherent in their language may go beyond the strictly immediate.

E.P. So, mankind would once again have to think about what precisely it is that differentiates us from all other animals.

P.T. Why must we believe that we humans are unique? Reminiscent of the search for El Dorado, this reminds me of discussions that abounded in England in 1860, the year after the publication of Charles Darwin's *Origin of Species*. A well-known English comparative anatomist of the nineteenth century, Sir Richard Owen, dissected many bodies, but he rejected some of Darwin's ideas, which were anathema to him. Owen said, "It is impossible that we descend from the monkeys because our brain is unique." They asked him: "How is it unique?" And he replied, "It has something called the hippocampus minor, a small entity beside the large hippocampus or hippocampus major. No chimpanzee, gorilla, or orangutan has a hippocampus minor." Then Thomas Henry Huxley appeared, who had begun to dissect the brains of chimpanzees, gorillas, and orangutans, and he said, quite rudely: "Look, here is the hippocampus minor of monkeys." So, that supposed singularity of the human brain was refuted. Sir Richard Owen's prestige began to wane.

E.P. Because of the hippocampus.

P.T. Because of the impact of the discovery that the hippocampus minor is not unique and therefore neither is the human brain. For Owen, it was a terrible, unacceptable revelation.

E.P. Ralph, so where does the human condition lie, if it exists?

R.Ho. It's always changing. If you observe any brain closely, you see that each is unique. Like each animal, natural selection has grouped different bundles of unique features together. Our brain is unique because of its unique combination: large size, certain bundles of fibers in the intermediate areas, and a certain reorganization of parts that are in the brains of all mammals. The human brain has no new differential characteristics, as poor Richard Owen discovered. The singularity of the brain and the behavior of the human animal is that it can impose an arbitrary form on its surroundings. No other animal can do that. I learned little by little as a child. I was born in 1935, and I lived through years of horror between 1939 and 1945, because of the rise and fall of the Third Reich. It made an incredible impression on me to realize that human beings were capable of dramatic response to arbitrary symbols. We create environments that so radically and suddenly changed the genetic groups and the frequency of genes in the world. Only the human animal does that, no other animal can. Even if a whale or a chimpanzee develops a concept of the future, we shouldn't fear anything—

E.P. But they should fear us! Ralph, you studied the brain and its origin, which dates from two million years ago. Do you think that the brain, as an instrument, can guide the hominids towards something better?

R.Ho. Personally, I am pessimistic. My view is as follows: If we examine the geological data on the evolution of almost all mammals and other animals, we see that there is a capacity of cohesion toward a morphological model that we call the genus. We are *Homo*, not primates, though we descend from them. This condition dates from about six million years ago. If we observe any other animal, its paleontological data extend back approximately between five and ten million years. Humans, as a genus, have only existed for six million years. Achieving evolutionary success, survival in the future, would take us at least another three million years only to achieve what almost all species of mammals already have. I wonder if we could imagine another three million years like these.

P.T. I'm optimistic.

E.P. And do you think that we can survive another three million years?

P.T. Not without great effort. We have developed a wonderful mechanism, the brain, but we don't always know how to use it in the most efficient way. This is one of our great problems, already present in today's education system. We must teach young people to "use your brains," as my old teacher always said, in the plural—"brains." Perhaps the great challenge of the future will be to develop the ability to obtain the most from what the evolutionary process has given us.

— 21 —

Immortality from Your Mother

Interview with Douglas Wallace

The energy generator of the cells is inherited from your mother, not from your father.
—Douglas Wallace

The other vital distinction in the biology of immortality we owe to the young North American scientist Douglas Wallace is not that Wallace discovered the existence of mitochondria in the cytoplasm of the cell—they had been known as "bioblast" or "cytodes" since the times of Richard Altmann, in Germany in 1890—but rather that, in a period that began in 1953, amid the tremendous hullabaloo about DNA and the promises implied by knowledge of its

Douglas Wallace is the Donald Bren Professor of Molecular Genetics and director of the Center for Molecular and Mitochondrial Medicine and Genetics at the University of California, Irvine.

sequencing, the young Wallace drew attention to the importance of mitochondria as generators of the energy the cell needs to survive and the poisoning that consumes it.

Douglas Wallace is the director of the Center for Molecular and Mitochondrial Medicine and Genetics at the University of California, Irvine, where he supervises work on the evolution of the mitochondrial DNA in human beings, thanks to which new data have been discovered about our longevity or our adaptation to environmental changes due to human migrations in our conquest of the planet.

Douglas Wallace: The mitochondria are the other human cells. Inside each cell, there is a colony of former bacteria of vital importance, the mitochondria, which are the generators of the cell's energy. They provide all the energy to the rest of the body. There are about ten to the seventeenth power mitochondria in your feet right now.

Eduardo Punset: And where do these mitochondria come from?

D.W. The mitochondria have a very interesting origin. In the beginning, about two billion years ago, they were free bacteria. Then they were surrounded by cells that finally gave rise to our cells and so began to live inside them.

E.P. In a symbiotic association?

D.W. Exactly. At first they gave protection to the nucleated host cell—which contained our nucleus—protection from the presence of increasing oxygen gas that appeared in the Earth's atmosphere. The Earth's atmosphere, in the beginning, had a very small amount of oxygen. With the invention of photosynthesis molecular oxygen began to be emitted as a waste product. This emission changed the atmosphere, from reducing to oxidizing, and many, many organisms died.

E.P. People forget that the bacteria changed the entire atmosphere of the planet.

D.W. The whole ecosystem.

E.P. Incredible, no?

D.W. Yes. These mitochondria had already invented the ability to take the oxygen from the air and react with the hydrogen of carbohydrates, fats, proteins, et cetera, from the food we eat, and to create something new. The energy that was released was now trapped so that it could be used by the cell. These mitochondria, since they contained oxygen, invented ways to get rid of the poisonous oxygen. Therefore, when the mitochondria were trapped inside the host cell, the first thing they did was to protect the host cell from the toxic oxygen, and this stabilized the symbiosis. Later, the host cell managed to induce the mitochondria to give it the ATP, the energy molecule. And the rest, as they say, is history.

E.P. And how do we arrive at the current situation?

D.W. The most incredible thing about the mitochondria is that, as they live in the cell cytoplasm, outside the nucleus, they are not inherited the same way as the genes of the nucleus. They are inherited through the cytoplasm. The magnificent human egg has a large cytoplasm that contains about one hundred thousand mitochondria. However, the sperm only has a few. Therefore, in fertilization, all mitochondria come from the mother, so the energy generator of the cells is inherited from the mother, not the father.

E.P. Energy is from the mother.

D.W. Without a doubt. The feminist movement wasn't wrong.

E.P. If the mitochondria do all these things, they must be very important for our health.

D.W. Of course. Mitochondria are like power plants that generate the energy of the city of Barcelona. The people are in the center of Barcelona, but in order to be able to use the energy, power plants are built on the outskirts, in the countryside. Installation for coal, nuclear energy, or hydroelectric provide the city with electricity. But those that generate energy and electricity burn fuel and produce smoke, like where the coal is burnt and energy is produced. The same happens with the mitochondria. Hundreds of mitochondria are in each cell and each one generates a little bit of energy for the cell. In the process they generate "smoke," in reality oxygen radicals. I'm sure you've heard of the oxygen radicals—

E.P. The free radicals—

D.W. The free radicals, as they are called, are a problem for health. The free radicals are like the smoke in the power plant. Over time, the free radicals produced by the mitochondria damage the DNA. When all the DNA of the mitochondria of the energy generation process is damaged, because mitochondria DNA can't repair it, the cell dies.

E.P. Doug, don't you think that thousands of millions of dollars are spent on finding out what happens inside the nucleus of a cell—I'm referring to the fight against neurogenerative diseases, for example—and, on the other hand, very little is invested in finding out what happens with the mitochondria, which, when it comes down to it, as you said, are what power the cell, keep it going?

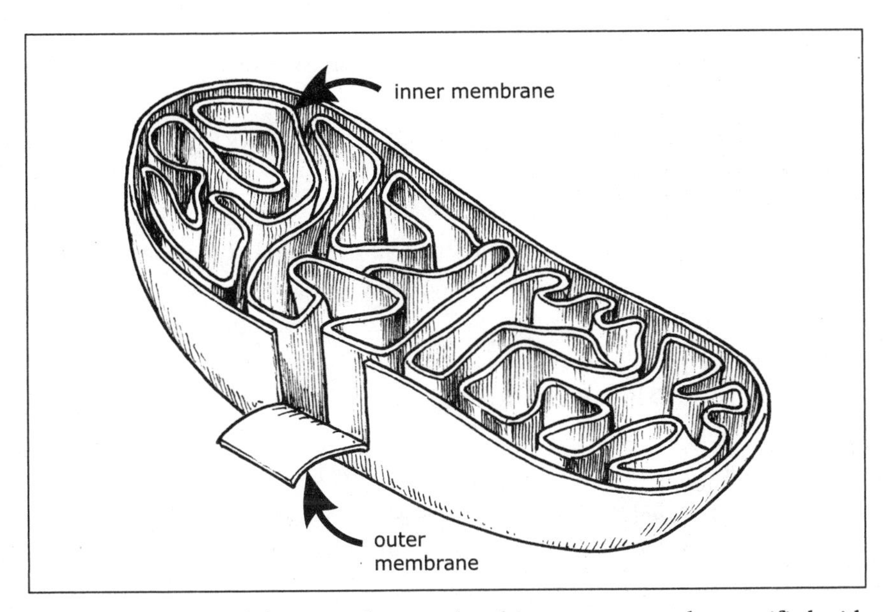

Fig. 4. Cross-sectional drawing of a mitochondrion as seen greatly magnified with an electron microscope.

D.W. Exactly. The human cell is made of two types of cell, and the nucleated part is only one. The other cell, equally important, is ignored. That cell is responsible for generating almost all the energy we use and, of course, for making the toxic oxygen radicals. Mitochondria, what's more, have another interesting property. Not only do they generate the energy and toxic waste, but also, when the mitochondria become ill, they have a self-destruction property that orders the mitochondria and the cell to self-destruct.

E.P. It's the famous apoptosis: a programmed suicide.

D.W. Exactly. Therefore, when the mitochondria's energy plants become useless, instead of having cells in the body that generate poisonous oxygen radicals, the mitochondria simply close down. They destroy cells they're in by apoptosis. In humans so many mitochondria are damaged at any one time that the whole cell dies. As they age, people have more and more cells with damaged mitochondria. Apoptosis occurs and the cell dies. In the end there are too few cells in the tissue to carry out its function.

E.P. And what can we do in the meantime? I mean before the mitochondria decide to apply the programmed suicide?

D.W. One of the things on which we are working is to try to develop drugs directed at the mitochondria that eliminate the toxic oxygen radicals. They resemble sweeps who remove the accumulation of soot. Therefore, if we reduce the contamination caused by mitochondria, we will be able to protect the integrity of the cell.

E.P. And what do you think of the pills sold in pharmacies to combat oxygen radicals? I refer to vitamins C, E, and carotene. Are they useful, while we can't yet directly affect mitochondria?

D.W. Yes. It's a start. Vitamin C, vitamin E, and carotene do reduce the effects of the oxygen radicals on the mitochondria. But they are not very efficient at elimination of the oxygen radicals. We develop drugs that will continuously, or catalytically, as it is called, reduce contamination, like catalytic converters in cars. The catalyst would be constantly regenerated and waste eliminated.

E.P. And if you are successful, what will aging be like?

D.W. Our hope is that we will prevent the progression of the degenerative diseases, which we believe are caused by damaged mitochondria: Alzheimer's, certain types of blindness and deafness, certain heart and kidney diseases, or diabetes. We believe that the damage caused to the mitochondria is the cause of the subsequent appearance of these diseases. Therefore, if we could make a drug to prevent this damage, we would be able to stop the problems that people develop as they grow old.

— 22 —

Inevitability of Aging?

Interview with Tom Kirkwood

We are not programmed to die.
—Tom Kirkwood

Our culture is the culture of unlearning. The river of knowledge has always been added to by tributaries that increase the available flow. Like evaporation in nature, some knowledge is banished into oblivion, lightening the available knowledge in order to be able to survive. But culture has always been the sum of added values: to the old religions are added new sectarianisms, to the most remote genetic knowledge are added age-old mutations, and, in particular from the sixteenth century on, the knowledge acquired generation after generation fills the planetary brain, libraries, and computers.

Tom Kirkwood is Professor of Medicine, Co-Director of the Institute for Ageing and Health at the University of Newcastle, and Director of the Centre for Integrated Systems Biology of Ageing and Nutrition.

From now on, and for several centuries, the metaphor of the processes of knowledge will be represented not so much by a river that flows into the sea, but rather by a stubborn salmon swimming back to its origins. We are experiencing a revolutionary educational change in which the structure of the new knowledge is dictated by the massive processes of unlearning. We will depend, more than on our knowledge, on our ability to unlearn.

Unlearning is applied to all the branches of knowledge, including scientific knowledge. The only difference between science and the other knowledge disciplines is that the raison d'être of scientific knowledge is precisely unlearning by means of experimentation and testing. In few knowledge disciplines is it as palpable as in the biology of immortality, the subject here. It is necessary to refute very deep-rooted myths in order to examine the processes of antiaging and regeneration of tissue.

The first great myth is the human certainty that we are programmed to die. Physical and mental health requires, first of all, the eradication of this fallacy, widely believed even among the scientific community that investigates antiaging. There are still too many scientists who are trying to find the internal clock that regulates aging. Tom Kirkwood, a gerontologist at the University of Newcastle in the United Kingdom, is a leading expert on genetics and the evolution of aging. He is a staunch opponent of the theory of programmed death. He maintains that aging is, in fact, the result of the accumulation of damage in the cells and tissues in the course of a lifetime.

Tom Kirkwood: There has been a lot of discussion about the clock that measures the time of life and destroys life when it runs out, but one of the great advances in the understanding of aging has been the finding that, in reality, there is no program for death. It is extraordinary that inherited knowledge leads toward survival, that is, precisely the opposite to what happens in practice, the death that ends all life. If we examine the body of a dying person, we see that all the cells and organs are trying to keep the body alive. The program that governs life never gives in to death.

Eduardo Punset: After the body has died certain cells continue to struggle to remain alive?

T.K. Yes. Because the message of death takes time to spread around the body. For minutes or hours after the death of the body there are still living cells. If we extract an organ from the body of a person who has just died in order to transplant it, the organ remains alive for some hours. The heart and the kidney remain alive, though the body is dead, and they can survive if they are put into another living body. Therefore, the cells of the organs fight for survival in a dead body. This is important to understand the process of aging, so we don't start by looking for a program that leads to death that, in reality, does not exist. Our genes, in part, determine our life. One of their functions is to ensure survival. The body ages and dies despite the fact that it is wonderfully programmed for survival, because it cannot survive indefinitely.

E.P. Unlearning myths, therefore, there is no biological limit on life.

T.K. The idea that there is a biological limit to human life is a great fallacy. It is true that, as we get older, it becomes increasingly difficult to survive for a long time. But it is like the world record for the mile: there is no absolute time that cannot be surpassed. For a long time, no one had managed to run a mile in under four minutes, and in the 1950s that was achieved. But, as the records reach ever higher, it is increasingly difficult to break them, though that does not mean that there is an absolute limit. And, in a way, the absence of an absolute limit on the duration of life is related to the absence of a program. As we understand more about the process of aging, we will reach old age in better conditions than previous generations—the human body's ability to survive has increased slightly. The world human longevity record is one hundred twenty-two years and five months. It will be a few years before it is broken, because no one is approaching that age, but, sooner or later, it will be broken.

E.P. You have suggested another mistaken but very deep-rooted concept. I think that it's not so much that longevity has increased as life expectancy, because fewer children die than in the past, right? Infant mortality has greatly decreased—

T.K. It's true. A hundred years ago, even in the richest countries in the world, life expectancy was very low, infant mortality was very high. Many children died before the age of two, three, or four years—and many by infectious diseases. If we calculate the average life expectancy between a child who dies at the age of four and an old man who dies at the age of eighty-four, we obtain an average of forty-four years. Fortunately, in the West we have now eradicated infant mortality to a great extent. It is extraordinary and we should celebrate it because it is the result of arduous human efforts.

E.P. Even so, we are a species that cannot avoid death. We get old and the wrinkles indicate the proximity of death. Why does it seem that certain species do not die or are immortal, whereas others die more quickly? Why is there such a great difference?

T.K. The length of human life is, at the most, one hundred years, whereas a mouse lives for three years, a dog fifteen, and a cat twenty-five. Investigating the reasons for these differences is a very interesting scientific challenge, related to evolutionary biology. We need to study how the process of natural selection acts on the genes of the organism and the design of the body, and the different aging properties it generated.

E.P. But why do species like the hydra or the anemone appear to be immortal?

T.K. Essentially, it is a question related to the fundamental properties of the organism, as the German naturalist August Weissmann discovered in 1880. Two types of cells exist in an organism. Those that create the new generation of offspring, the reproductive cells, of the body—

E.P. The germ cells, right?

T.K. Exactly, the germ cells. And the other cells, those that form the brain, the heart, the kidney, the skin, do not contribute genes to future generations. These disposable cells are called soma. Aging is a property of the ordinary body parts, the soma. The germ cells are extraordinary. When a sperm from the father joins an egg from the mother, the first cell of the new body is formed. It begins to divide. But if we go back in time and we ask ourselves how the father and mother's sperm and egg came about, we will come to—

E.P. Three billion years ago—

T.K. Exactly, three billion years ago. Therefore, we are a biological miracle, because we are the product of an uninterrupted chain of cell division that has lasted three billion years. So the reproductive cells, germ cells, are therefore immortal. They are transmitted from one generation to another. And these reproductive cells manage to avoid aging. Organisms like the hydra or the sea anemone seem immortal because, in reality, the entire organism is germinal. If a piece is cut off a hydra and put in a clean glass of water, it will develop until it becomes a new hydra, because the germ line is distributed throughout the organism. However, very few of the cells of humans are germinal. If you put a piece of skin into a cup of water, it will not germinate a new human being, because the cells of the skin are somatic. Aging is a property of the soma.

E.P. While you were speaking about immortal cells, germ cells in the human body, I remembered that physicists argue that atoms are eternal, and that 99 percent of our body is composed of atoms. That is the paradox: human beings are condemned to death though we are composed of immortal cells and almost immortal atoms. You believe that finite or disposable soma is the main reason we age? What do you mean? Reading your research, it seems that you conceive of the genes as very intelligent entities that carry out a cost-benefit analysis, and based on this, they decide whether or not to invest their energy in maintaining or prolonging life. According to your theory, in that decision lies the aging, caused by a lack of maintenance.

T.K. Of course. I want to stress that the genes don't think, though sometimes we attribute a consciousness to them that they don't have. Genes don't make decisions. The genes that have better strategies due to mutations and chance impose themselves among the different forms of genes that exist among the population. But sometimes it is easier to talk about the genes as if they could think, as if they were aware. Let's continue with this language. Let's try to understand what determines whether or not a gene is successful. The success of a gene is measured by the role it plays in the construction of a body that generates more copies of the gene in future generations. The genes direct the construction of the body and everything the body does in the course of its life. Above all, the body has to grow and undergo a complex development process, because life begins with a single cell.

E.P. Until it reaches trillions of cells.

T.K. Yes, thousands and thousands of millions of cells. While the growth process develops, mistakes are made. Every time a cell divides, mistakes are made in the way it is copied. And every minute of life proteins are damaged, simply due to the vibrations that exist in a warm organism. The proteins are shaking very fast and when new proteins are generated mistakes occur, so things inside go badly even when we just sit here and talk. Many things go wrong in the body. The fact that we remain alive week after week, month after month, year after year, is the result of a tremendous investment in maintenance. The body's work to keep the cells in good condition is extraordinary.

E.P. You have written that every day each cell receives ten thousands knocks. It's incredible, ten thousand knocks!

T.K. Yes, we have millions of millions of cells. Each one of those cells of the body receives about ten thousand knocks in DNA every day. These knocks come, surprisingly, from the oxygen, a friend of ours. But that's another story. Sometimes we forget that oxygen is a friend that provides us with the means to be able to live but—

E.P. But it is a murderer.

T.K. It is both friend and killer. The most important question is that if the cells tolerate so many mistakes and knocks every day, we feel lucky to survive one week. The reason why we can survive more than one week is because we have a magnificent DNA repair system, which recognizes damage and works to repair it. So, of the 10,000 knocks received in the DNA of one of your cells, tomorrow 9,997 will have been repaired. But it is not done gratuitously. The genes have to decide how much energy they invest in the repair and maintenance processes. Summing up, we have already spoken about growth and now we should focus on repair and maintenance, because the genes have other things on which to invest the energy and resources, such as making babies. It is essential that the genes make copies for future generations.

E.P. You suggest that while the genes are busy making babies they cannot carry out maintenance and repair processes.

T.K. It's like the money in a current account. If you decide to go on holiday or dine in a fancy restaurant, you won't have money in the bank to repair the car the week after. Money can't be spent twice. An organism has to decide how much it will spend on the maintenance of the body. The real world of humans is very singular and it hasn't always been like it is now. We have evolved until we have become very intelligent and live for a long time, we don't die by being devoured by predators, and today virtually no one in the West dies from infectious diseases. But two centuries ago, life was different. People died young and life expectancy was not more than thirty or thirty-five years. But if we imagine the genes in the context of a life of about thirty years, how much do you think the genes would invest in the maintenance of the body?

E.P. Very little. Because life would end very quickly.

T.K. Very little. For such a short life a perfect body is not necessary, it's a waste of energy. Unfortunately for us, the genes treat the body like disposable soma. Once the continuity of the next generation has been guaranteed, by means of new births, we get old. It's a luxury, a loss for the genes to invest in the maintenance of the body once it has reproduced. We have evolved to have limited maintenance of the cells in the organs of the body.

E.P. I would like to ask you two questions. The first is that, as we now live many more years, the investment in maintenance should increase, right? And the second thing is whether aging is related to the fact that the cells do not continue dividing after fifty or sixty years. Why on earth is it that way?

T.K. Forty years ago, Leonard Hayflick did a study that gave its name to the phenomenon. The Hayflick phenomenon is that, as you said, the body cells, the somatic cells in the laboratory, can only divide fifty times. These fifty cell divisions are enough for the cells necessary for a lifetime. If we take the number two—because the cells divide in two—and multiply it by two, and so on successively, each multiplication simulates a cell division, as each division doubles the number of cells. When we have multiplied two by two, the result by two, and so on successively fifty times, the number of cells we obtain is greater than the number of cells that the human body needs in the course of its life. On the other hand, in a laboratory cell culture the cells age and die due to an accumulation of errors. Now people investigate the nature of the mistakes that are made. They are the beginnings of the finite soma theory explained at the level of the individual cell.

E.P. Because what happens in the body can be analyzed in a cell culture—

T.K. Yes, but not all the cells of the body can be cultivated in this way. Some cells of the animal don't divide during our lives, especially those of the brain, which can live for over a hundred years. We know this because people whose brains continued to work properly have lived over one hundred years. In these cases, the maintenance of the brain cells is necessary and there are extraordinary repair processes. The nature of the proteins changes, as they might have suffered damage, perhaps from oxygen. The cells have waste disposal systems to eliminate the harmful proteins, to repair the DNA. Even though the cell doesn't divide, the DNA continues to suffer, so it doesn't matter whether it's a cell that divides or a cell that hasn't divided since we were babies, like those of the brain, the cells need maintenance systems that keep them healthy for a time. Unfortunately not forever; as we grow older the brain cells start to lose functions.

E.P. Let's try to analyze these threats from the point of view of the average man and woman. The first threat is mutation. Because of radiation, the cells mutate and sometimes they mutate erroneously, and can't repair themselves. The first great threat, therefore, is mutation.

T.K. That is what happens with cancer. As we get older, there is a greater risk of developing cancer, and the majority of cancers are caused by mutations. We know where the mutations and the mistakes made by the cells when they copy themselves come from.

E.P. The second threat is the so-called free radicals, the destructive beasts that act against the cells. What on earth do these free radicals do?

T.K. The free radicals are the negative part of our friend oxygen. The body uses oxygen by breathing in a mouthful of air, which is transmitted to the blood and this carries it to every cell in the body. Once in the cell, the oxygen penetrates inside it by means of small organelles, the mitochondria [chapter 21]. These are capsules on the inside of the cells where, in reality, the cell burns the oxygen to create energy. Therefore, this oxygen is a fundamental part of the generation of energy inside the cell. But you know what happens if you use oxygen in the fireplace in your house in order to create energy. The majority of the fire remains inside the fireplace, but occasionally the fire throws out sparks and it's dangerous, as the fire might spread outside the fireplace. Well, the same thing happens inside the cells of the body: the majority of the oxygen is used in a safe way, inside the mitochondria, but it is estimated that 2 or 3 percent of the oxygen molecules escape from the appropriate chemical channels in which the process is carried out and—

E.P. And wreak havoc.

T.K. And wreak havoc, because oxygen is a highly reactive gas. We have all seen how oxygen rusts the metal of our car, or what oxygen does to butter if we leave it exposed to air. Oxygen is very harmful and can cause a lot of damage inside cells. The free radicals from the oxygen attack the first thing they find, and if it is the DNA, they damage the DNA. If it's a membrane, they damage the membrane. If it's a protein they destroy it. And that is happening every minute of your life.

E.P. What remedies are there? We have already identified the threats: mutations, oxidization, and the fragility of the mitochondria and the DNA of the mitochondria. It seems that there are only two strategies. The first one consists of reducing exposure to these daily attacks—you will tell us how—and the second one is to improve somatic maintenance.

T.K. It is essential to understand what happens in aging in order to be able to combat it in time and increase the likelihood of reaching old age in good shape. There are two ways to solve aging. If it is due to an accumulation of damage to the cells, we can try to protect ourselves from the damage. The other is to try to improve the repair processes. In fact, we can do both.

E.P. Such as what?

T.K. Aging is the result of damage, and one of the ways in which this damage happens, and which can be reduced, is from what we eat. We are what we eat; almost everything we ingest forms part of the body. If we eat poorly, for example, excessive fat or sugar, it is harmful for the cells and tissues of the body. It is not good to have too much sugar circulating around the blood because it damages the proteins and causes many problems. That is why diabetics have so many problems, and before modern treatment with insulin, diabetes was considered an illness of premature aging. We should not expose the body to excessive damage, and of course we all know what happens to smokers.

E.P. Before going on to the subject of tobacco, explain to me what is revealed by the experiments in which the calorie intake of rats was reduced by 30 percent, with the result that they lived longer lives. Is it applicable to human beings?

T.K. I think it's a depressing prospect, to have to reduce our food intake by 30 percent. And it has not yet been demonstrated whether it works with human beings. But it does work with rats.

PART THREE

LIFE ON AN ANIMATE PLANET

Bygone Biospheres

INTRODUCTION

Earth, were it like its neighbors, Venus, closer to the Sun, and Mars, farther away from it than us, would be dry. The air would have less nitrogen, no oxygen, and would be loaded with carbon dioxide. But here on Earth we have green life and blue skies. Life has been the most important surface geological force of any other in the last three billion years. Because life drew down the carbon dioxide into limestone rocks the Earth was cooled. Our planet Earth retained an average over her surface of 3,000 meters (3 kilometers) of water. Mars and Venus have an average depth of surface water far less than even a single meter. Mars is freezing cold and Venus is boiling hot. They are both far drier than bone. Their worlds are far more acid than our own; Venus's clouds are composed of sulfuric acidic, for example. James Lovelock's Gaia hypothesis explains why, unlike our neighbors in the solar system, even beyond Venus and Mars, Earth is just right.

We live in a physicochemical world but life stretches some of the physicochemical rules. Life here extends from the equator to the poles, from the abyss to the mountaintops and in the deserts and deep sea vents. Life is a planetary phenomenon here. Life was teeming on Earth for 3 billion years before we talking apes ever entered on the scene. If you understand in what sense Ken Nealson means that "life is a mistake" you will also see by what criteria we might infer the existence of some kind of life on other planets.

— 23 —

Life, Master of the Earth

Interview with James E. Lovelock

"We will suffer for all we have done and only then will we understand how wonderful our planet was.
—JAMES LOVELOCK

Eduardo Punset: Where do we start? You say: "Life is in charge of everything and controls evolution" and I think of that wonderful photograph of Earth seen from space. We will talk of Gaia later. I see that vision of a lonely, blue planet, a vision that changed our perception about both the place where we are and what we are.

James E. Lovelock: That was perhaps the most extraordinary image I have

---

James Lovelock developed the concept that is now known as Gaia Theory and pursues independent research at the Coombe Mill Experimental Station in Cornwall, England.

ever seen: the picture of Earth from space. But I believe that the true view in the mind of the astronauts when they looked at their planet was different. When you talked to them, they always commented on what they were excited about, how they realized that the blue sphere was home, their home, even more than the street in which they lived, or their city, or the country of which they were citizens.

E.P. It is a wonderful picture.

J.E.L. Indeed.

E.P. Did you get to know John Oró, the Spanish scientist who worked at NASA for many years?

J.E.L. Yes, of course. We worked at the same university in Houston.

E.P. He too was impressed by this view of the Earth from space. I remember when we met in Barcelona, at the first parliament ceremony, when democracy was restored in Catalunya after Franco's death. Both of us were newcomers; he from Texas and I from the International Monetary Fund in Washington. We had been attracted by the prospect of seeing democracy in Spain at last. "Now I am convinced that things will change," he told me. "When people look at Earth from the outside, something strange, revolutionary will happen: people will alter their thinking." "Are you sure, John?" I asked him. I have not seen those changes yet. Have you, James?

J.E.L. Those changes have not yet taken place, and I think this is because we have not yet done enough harm to the Earth seen in that photograph. We have not yet realized how wonderful a place it is. In the next century we will pay for our actions: the destruction of the atmosphere, the removal of natural habitats. We will suffer for all we have done and only then will we understand how wonderful our planet was.

E.P. When I met John Oró—we were like two Martians in a country that suddenly had become ecstatic—I tried to convince him that borders would soon disappear. "From space it is clear that there are no borders or separate countries, and from now on there will no longer be borders," I told him. Nevertheless, there are still borders everywhere, even in my small country, Catalunya.

J.E.L. As Edward O. Wilson, the Harvard scientist whom you also interviewed [see chapter 5], has asserted, the problem with human beings is that we are tribal cannibals. It is impossible to change that, unless perhaps we are modified by genetic engineering.

E.P. It is incredible, the slowness of cultural change. I am really surprised whenever I hear people say that everything is changing very quickly. From your perspective as an independent scientist, do you see changes at all?

J.E.L. It is necessary to change for everything to remain unchanged. I think that all is tribal. It does not matter what kind of system is ruling—capitalist, religious, or anything else—there is always a hierarchy with a tribal leader, like in primitive tribes, and we cannot change that. It is part of our genes, of ourselves. It will never change, we will always have tribal systems.

E.P. As happens with chimpanzees.

J.E.L. Exactly. We look so much like them! There are only differences of grade.

E.P. Something has always intrigued me. There is a kind of consensus in the scientific community, mainly among physicists but not only among them—I am thinking of people such as Paul Davies [see chapter 36], for example—about the fact that we are lucky because we live in a favorable universe. We appeared in the right place at the right moment. If there had been even very small deviations in the profile of the four forces that rule the universe, for example, there would be no life. On the other hand, there are also scientists like yourself that do not share this opinion. "No, no, it is not that," you have said, "we are in the right place just because life exists; it is life that has modeled this lovely planet that it is so favorable to life."

J.E.L. Yes, Eduardo, this is so. Classically scientists have said, "Fortunately, Earth is at the right distance from the Sun, and the temperature is just right for life." But this is nonsense. Perhaps there was a moment—maybe when life first appeared on Earth—at which Earth was more or less in the right place; but once life originated on our planet, the planet no longer evolved as its neighbors did. We never lost water and became desertified, as happened to both Mars or Venus. In some way, life is in charge of everything and controls evolution. Both evolutionary systems, the inorganic and the living ones, run together. For this reason the planet is always a good place for life to exist at any moment.

E.P. It is possible that other forms of life may have existed that might have competed with it.

J.E.L. Yes, it was so in the past. When bacteria were the only inhabitants of Earth. For nearly one billion years, the atmosphere was dominated by only one reactive gas, methane, and there was hardly any oxygen in the atmosphere, almost none. We people would have found that world very unpleasant and probably stinking. Yet life evolved and provided us with an atmosphere with oxygen similar to the one we have now. Things changed.

E.P. So, it is not true—as often has been said—that the introduction of oxygen was a real cataclysm for species on the planet. Reactive oxygen was salvation for complex life.

J.E.L. I think that I am to blame for having suggested, some thirty years ago, that oxygen was the atmosphere's greatest pollutant ever known and that it exterminated a countless number of species. I realize now that I made a mistake because, if we think of it, one of the first kind of organisms on Earth, the cyanobacteria, were photosynthesizers. This must have been so because the first energy source was the Sun and something had to invent the process to obtain energy from the Sun in an efficient way. When cyanobacteria were able to adapt, they started to generate oxygen inside their own cells. But oxygen is a poisonous, destructive substance and chemically is very active. For this reason, bacteria were forced to come up with some system to free themselves from poisonous oxygen. They invented an enzyme, superoxide dismutase, which splits doublet oxygen. So when oxygen appeared in the atmosphere, at the end of the Archean, cyanobacteria were used to oxygen and knew how to manage it. Oxygen did not kill them. It is true that some organisms did not feel comfortable with oxygen, but they decided to live happily underground. Now we have 21 percent oxygen in the atmosphere but, nevertheless, anaerobic organisms now probably outnumber those that were present in the Archean. They are inside us, in our guts, and beneath all underground layers of the Earth.

E.P. And they live with no oxygen.

J.E.L. They live very happily because others produce food for them.*

E.P. Certainly without oxygen there would be no ozone, and without ozone the levels of radiation coming from space would be higher.

J.E.L. In my opinion, this is an exaggeration. Before the appearance of oxygen there might not have been ozone and yet there was life. Surely what happened was that, before oxygen, there were decomposing methane products, which formed a sort of fog in the higher atmosphere that filtered ultraviolet rays, as ozone does today. When oxygen first appeared—even in

*Anaerobic microbes that produce food for themselves, probably an Archean eon legacy, are still about in lakes, muds, and well-lit ocean water. These include photoautotrophs (many different purple and green photosynthetic sulfur bacteria, e.g., *Chloroflexis*, *Heliobacterium*) and chemoautotrophs that produce food but no oxygen. They live in unlit zones in lakes, oceans, and rocks. Included in this group are methanogens. Others use hydrogen sulfide (H_2S) or iron as their source of energy, which they oxidize. They make their body parts from carbon dioxide. They eat no organic matter (no food). All are bacteria; people tend not to know about them.—LM the absence of enough ozone and methane to prevent the absorbtion of more ultraviolet radiation—life continued. I do not believe that ultraviolet radiation was as harmful as is touted. The difference in the amount of ultraviolet radiation between England or Spain and the highlands of Africa, such as Kenya, is almost eightfold. Radiation is eightfold more intense in Kenya. Nobody has ever heard of treatment needed there for solar burns. Nobody. Life goes on. It is easy to become habituated.

E.P. So we may think of a disaster, such as the destruction of the ozone layer, but the remedies may be supplied in other ways. Earth will find a way out?

J.E.L. Earth, yes. The problem is that, if the ozone layer disappeared, light-skinned humans would be the most affected. The rest of life on the planet would quickly adapt to the new circumstances, but we have very transparent skin and thus we are very sensitive to solar burning.

E.P. comments: I do not think that in the history of evolution a more dramatic and unique event has ever happened as the oxygenation of the entire atmosphere by growth from one type of microbe; the cyanobacteria. When you see them, so elegant in their green dresses, gliding in their aquatic habitats, it is hard to believe that we owe them our life, or, what is the same, the air we breathe. Until they finished oxygenating up to 21 percent of the atmosphere, the rest of us could not venture forth. Cyanobacteria are the first credible witnesses that—as Lovelock suggested more than thirty years ago—life is in charge of everything and controls evolution.

Another example is soil worms. We would have no agricultural products nor sustainable agriculture if they had not dug up the soils thousands of years ago. They were the most efficient plow, and were present long before the Romans invented this tool. Their role in the aeration of soil was crucial, as Charles Darwin suggested in the last book he wrote, published in 1881, *The Formation of Vegetable Mould Through the Action of Worms with Observations on Their Habits*. Darwin estimated that an average of six to eight tons of dry soil circulate through the guts of those small animals yearly. They carry potassium toward the surface and phosphate to the subsurface, and they add nitrogenous products from their metabolism. But the importance of worms in the sustainability of cultivated fields had already been recognized before Darwin. In ancient Egypt, where survival closely depended on the fertility of the Nile Valley, the pharaohs punished those who harmed worms. Aristotle referred to them as "the soil's intestines."

The worms' activity allows better absorption of rain so the amount of underground water increases. Soils retain humidity for longer periods of time, which allows greater variety in the types of crops. The planting of trees makes possible the reduction in the soil temperature between ten and fifteen degrees. All this favors the presence of bacteria and other organisms that complete the cycles of organic-matter decomposition. The crop-growing areas remain in good condition, soils are less compact, and they retain more mineral nutrients. Beyond soil preservation, the activity of worms can transform an arid and almost barren zone into a rich grassland within a period of thirty years.

E.P. The astronauts saw Earth from the space with their very own eyes, and I saw their photographs. How were you able to see it?

J.E.L. I was very lucky. At that time I worked at the Jet Propulsion Laboratory in California. I was able to observe Earth, Mars, and Venus through very special scientific eyes: infrared telescopes. I obtained data about the composition of the atmospheres of these three planets and could realize the extraordinary differences that exist between our planet, Venus, and Mars. The atmospheres of Venus and Mars are replete with carbon dioxide and only a few other gases in close to what chemists call an equilibrium state. I could thus predict that those planets had no life. By contrast, Earth has an atmosphere where combustible gases, such as methane, mix with oxygen. Ours is almost an inflammable atmosphere. If it were a bit different, it would explode. Ours is also a very fragile atmosphere. However, it has persisted for billions of years. How is that possible? This question crosses my mind again and again. Something is required on Earth that regulates the atmosphere and keeps the gas composition nearly constant.

E.P. What is fascinating is that so much depends only on this thin layer of the atmosphere, characteristic of this place.

J.E.L. The same occurs when you look at a person. When I look at you, I see only your face. All the internal mechanisms of your body are invisible to my eyes. The same happens with Earth: the atmosphere is what is seen.

E.P. If Mars is really different, then I understand why, more than thirty years ago, you said that it was not necessary to send robots there in order to know whether there was water or life. You only had to observe its atmosphere and realize that it was in a state of balance that vastly differed from Earth's atmosphere.

J.E.L. You are right. But one always learns with time. Now I would say that Earth is a living planet, living, in the sense that it regulates itself, like you, me, or any animal. Mars is a dead planet, it is like the body of someone that stopped existing, someone who died or never was alive.

E.P. The concept of life is very important to all of us. But when you say that the planet is alive, this means that it is even more important to you. You must have a more precise idea of what life is!

J.E.L. I am a scientist, but you know that we too belong to distinct tribes: biologists, physicists, chemists, et cetera. If you ask every one of them, each will answer you in a different way, his way. The biologist will say that life is something that reproduces itself and that the mistakes in replication are corrected by natural selection; this seems to be the only thing that most biologists want to know about life. The chemist will say that life is something that metabolizes, that takes chemical elements from the environment, processes them, and, in a specific way, chemically transformed, gives them back. A physicist will not say anything like this. Life is a system that works like a refrigerator: it takes free energy and constructs itself in the shape of a dissipated structure.

E.P. But how did all that start? From the inorganic, from the immaterial? Or from crystals that observed themselves and then began to replicate? You said before that the first cyanobacterium already had accurate instructions that enabled survival in its surroundings. Who gave it these first instructions?

J.E.L. I have no idea. I do not think that anybody else knows, either. It is still one of those big questions for which we have no answer yet. My friends from the scientific, academic world like to speculate about what existed before the big bang, or about how life developed on Earth in early times. Up to now we do not know, we only know that—and this is very important—we are very lucky because we are part of a self-organizing universe. Wherever there are energy fluxes, like solar light that comes in from a star, systems and structures form that survive for some time and are then extinguished. Life is one of those systems, but distinct from the rest because it is almost immortal. The reason that life is almost immortal is that it can transmit the knowledge of what must be done to sustain it from one generation to the following.

E.P. If something catastrophic happens, and we lose this faculty to transmit genetic knowledge to others, do you believe that everything would still start again?

J.E.L. I think that it is too late for this planet.* Gaia has existed for three or four billion years, but it has been calculated that Earth hardly has one billion

*He probably means too late for humans on this planet.—LM

years left before it dies.* The Earth is an old lady, like me. I am almost eighty! We can still be annoying, and we are beginning to deteriorate. This must not be forgotten.

E.P. We can, then, talk about the old age of the planet. Have we gone beyond the dividing line between youth and old age? Have we already crossed it? Has Gaia crossed the threshold?

J.E.L. Yes. We talk about an old lady who deserves respect.

E.P. We are so ignorant and careless as we fill the Earth with acid rain. We provoke warming of the planet through the greenhouse effect. Thermal changes imply an increase in the level of the sea and the disappearance of coastal towns. We contemplate, undaunted, how the hole in the ozone layer grows. Carbon dioxide concentration increases in the atmosphere. We no longer have the excuse of not knowing why Mars died, or never lived. Everybody agrees that terrible things can happen if we do not change. Let us assume for a moment that you are wrong and it is not too late. Is it possible to do something?

J.E.L. We can do something if we understand that the Earth is a living, habitable planet. It is not possible to farm all the soil to feed people. A great part of the Earth's surface is needed to protect the environment and to maintain the air in an appropriate form for us to breathe. The destruction of the Amazon rainforest affects the climate and the well-being of the entire world. We only think of humanity. When we begin to realize that the Earth is a living planet, we will know that we cannot discount all other mammals! We live in a century in which human rights have been the concern of focus. We tout that the most important things benefit humanity. I say no. This approach is erroneous. We should first worry about the Earth because we are part of it, we totally depend on it. If we do not do this, humanity as a whole will suffer.

E.P. Tell me, please, specifically. What we can do to prevent the Earth's demise?

J.E.L. We can obtain much more food directly from the chemical and biochemical industries. We need the soil only in a very restricted way in order to produce food. The great rainforests of the planet might be recovered so that they resume their function: to regenerate the air. Thus we might allow the planet to again become a suitable place for life. There are ways out. But they are based on technology, and not its abandonment. Some environmentalists are wrong when they say that problems will not be solved until we stop driving

*This estimate is likely to be too short. In fewer than five billion years the sun's radius is expected to be here, at the Earth's orbit. It is unlikely that any life will survive. Who knows?—LM

cars, burning fuel, or eating transgenic food. Imagine a group of environmentalists flying over the Atlantic on a jumbo jet and, out of the blue, they notice that they are expelling a huge amount of carbon dioxide into the atmosphere. The passengers decide to send a delegation to the pilot and they tell him to switch off the engine and let the airplane glide. Surely this will not work!

E.P. comments: James Lovelock is—along with Edward O. Wilson—the conservationist and environmentalist most emblematic of a generation of precursor scientists. Both were born about 1920 and have entered their eighties, both have challenged the scientific consensus of their generation. Both were excluded for some time, but neither should be underestimated. They are two exceptional, rigorous scientific minds. Both enjoy numerous followers, often more from outside the academic community than inside. Many of the followers were and are young people, who understand their intentions and share their fierce independence. Many know hardly anything of the methods and logic of their thoughts. "If we are on a jumbo jet, we need fuel in order not to fall. . . ." There are no simplistic, or even simple solutions to explain a phenomenon as complex as the persistence of airplanes in flight or, for that matter, the persistence of life. Both Lovelock and Wilson are convinced that, socially, human tribal behaviors and organizations prevail. Both of them, for this reason, will likely continue to be exiled colossi.

J.E.L. I think from a practical point of view, that most importantly we need to overcome our fears about nuclear power [Lovelock, 2006]. I understand that, while we have good reasons to be afraid of a nuclear war, it would not be so destructive to civilization. But nuclear power is good. It is the only energy source that does not harm the atmosphere. It does not cause damage. Nuclear reactions only involve threats to people, our pets, and other food mammals. But not to the Earth. If we went back two hundred years ago, when we were only one billion people, we might have saved ourselves with renewable energies: organic agriculture, alternative medicine, and all the rest. We might have done whatever we wanted, but now we must pay the price for having exceeded six billion people. We exert so much pressure on the Earth that we are forced to use technology to feed and maintain ourselves. I give you an example of what can be done and of what is done incorrectly. When I moved to this house [Coombe Mill in Cornwall, on the Devon border, near the town of Saint Giles-on-the-Heath and the small city of Launceston] twenty-seven years ago, this river was so full of salmon and trout that we hired a warden to prevent Sunday hikers from fishing. But environmentalists came, and said to the farmers: "Do not put nitrates in your gardens because nitrates harm people." So the farmers decided to replace nitrates with manure, and when manures and purines covered the fields, the rain dragged them into the river, and the fish died. Today, the river hardly has any fish or alga life. The problem with the ideas of the Greens is that they are well meaning, but they do not know how to evaluate or assess the consequences of their emotional recommendations.

E.P. Let us discuss your chemical production response to demographic growth now. The problem with pollution and waste is probably inseparable from human existence. But not for the planet. The Earth can solve it. So, let us resort to nuclear power in the near future. What else?

J.E.L. It will be necessary to impregnate ourselves with a new spirit. During times of war, tribes join together and make great sacrifices. They even offer their lives. If people thought of the Earth as their home, which it is, and the fact that it is in danger, maybe they would behave with similar good sense. Things can begin to look badly; just imagine that a city such as London is flooded and becomes uninhabitable owing to sea-level rise. Were we to stop burning fossil fuels right now, the sea would go on rising for another fifty years. But it would not increase as much as it is doing now. Maybe then we would have time enough to move back from the coastal areas. People would join together as during times of war. We would all accept the need to make sacrifices for the sake of our community. Now it is useless to advise people not to drive their cars because it is harmful to the environment; they believe that their jobs are more important. In times of war, things are different.

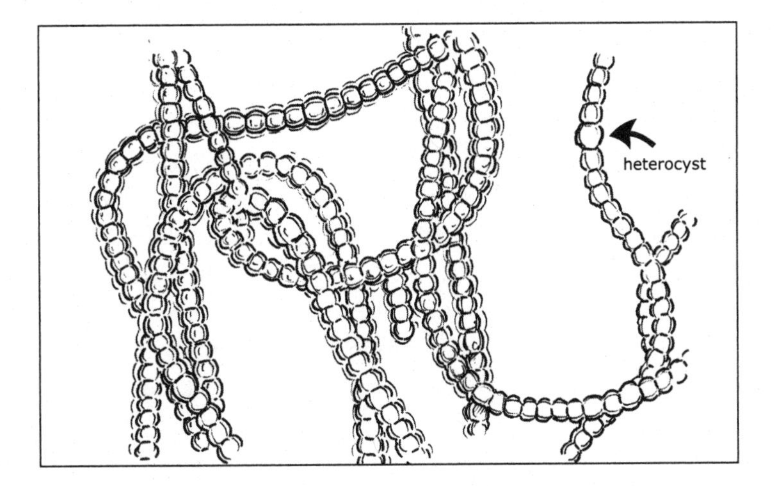

Fig. 5. The filamentous cyanobacterium, *Nostoc.*

— 24 —

Life Is a Mistake

Interview with Kenneth H. Nealson

Let me give you this rock and ask you:
Was there life on Earth at one time?
—Kenneth H. Nealson

As a researcher at NASA's Jet Propulsion Laboratory in Pasadena, California, Nealson founded the first Astrobiology Group. The charge is to develop general ways of detection of life and other extraterrestrial life-forms about which nothing is known in detail, both on Earth and beyond. By trade a microbiologist, he continues to lead an interdisciplinary research team and teaches now at the University of Southern California in Los Angeles.

The challenge to Nealson's team, to design a manual for finding life in

---

Kenneth Nealson is the Wrigley Chair in Environmental Studies and Professor of Earth Sciences and Biological Sciences at the University of Southern California.

space, has led the search for chemical clues for the presence of life. Life's stubborn refusal to immediately obey rules of chemistry and physics means that its inevitable waste products (spent gases, mineral deposits, acidification of surrounding waters because of acid end products, and so forth) reveal its activities. Not a matter of investigation from bottom to top (soil to satellite) nor from top to bottom (aerial to microscopic view) but rather of chemical identification of "what is there that should not be there." The only certainty is that life proceeds by energy use and matter incorporation. Solar or chemical energy is always required to flow to form cells to maintain them, to make the complex mixtures of interacting chemicals that is flow, growth. The course of the universe toward disorder and decrease in temperature is locally diverted in living systems. As he says, "We should not be walking on this planet, as all the laws of chemistry impede that this should happen randomly." If life is there, where it should not be, he is determined to detect it by its activities, undeterred and unobstructed by preconceptions about animals, plants, or even bacteria—familiar forms of life on planet Earth.

Kenneth Nealson: We should not be walking on this planet, as all the laws of chemistry imply that this should not happen randomly. Let me give you this rock and ask you a question. Was there life on Earth at one time?

Eduardo Punset: I suppose the rock comes from this planet—

K.N. Yes, it is an Earth rock. Now your task is to tell me if life exists or did exist on Earth in the past only by analysis of this rock. You may not examine any detailed chemistry. If life's chemisty was in the rock it is now gone. The DNA, the RNA, the proteins or any other type of molecule that we know exist in living beings is not available anymore. You must take a different approach. And that is what I call a non-Earthcentric approach to life detection.

E.P. What criteria then will you use for life detection—if not organic chemistry? Locomotion? Movement?

K.N. No, because this rock is two and a half billion years old. Nothing is moving. It does not now contain anything that is alive.

E.P. Will you check for the presence of water?

K.N. No. There is no water or not much. None bound in the rock is really relevant. Nevertheless, there are here all the indications of life. In fact, we have already shown statistically that life can be inferred from rocks like this! Now, we have to carry out the investigations by use of the right instruments. The fundamental idea is that all life must have some kind of structure that allows the conversions of one type of energy into another type. Life transforms energy but cannot do it without some sort of battery, transistor, or other energy converter—some kind of energy source that left clues.

E.P. Transistor? Battery?

K.N. I mean an energy transformer. Life is a system. First we look for certain structures, but structures alone are not clues enough. The structures must have a chemical composition. The chemical compounds should be the kind that by chemistry alone should not be there. Because life is what many people call negative entropy. This is something that is—

E.P. Not very common.

K.N. Something chemical that, given the details of the planetary environment, should not exist in that place under those conditions. Chemicals are there that should not be there. Life is always a way of taking energy and creating structure made from chemical products that differentiate it from what is simply in the rock. And in fact, if you ask the question about life on Earth, you will discover that carbon, phosphorus, sulfur, nitrogen, and a few other metals all together, are present in certain proportions every time there is life. And when there is no life, these elements are not there. They are certainly not there in specific proportion. It is a perfect diagnosis. Therefore, in order to analyze a rock like the one I show you, we must develop methods for measuring chemical products on a microscale, and raise a flag every time we find these combinations of essential elements derived from life's molecules, that should not be there. We look for a structure in rocks but made up of the wrong components!

E.P. Are you saying that life is made up of what should not be?

K.N. Life is a mistake! It should not exist. According to standard chemical equations, it should never exist, and the reason for this is that life is very complex. It requires a lot of energy—a continuous input, a flow, of energy. Structured energy flow, sources of energy flow within certain limits is an absolute requirement for maintenance of life. In other words, low-entropy energy, in order to achieve something like life or this rock as remnant of life, must have been here. We cannot make life through any sort of chemical process at random. We have to realize that what we seek is something halfway between thermodynamics and structure [see chapter 21].

E.P. Is it true that a bacterium inside an insect trapped in a piece of amber from ten million years ago, found in the Dominican Republic, was activated? Could this bacterium have hibernated in the amber for ten million years?

K.N. It is true. It is a bacterium belonging to the bacilli group that have spores—

E.P. Like the *Bacillus subtilis*—

K.N. Yes, very similar to *Bacillus subtilis*; the amber bacteria belong to the same group. You can find *Bacillus* spores in the stomach of an insect, extract these, and put them on a slide. What you find is that they are the same bacteria found in the stomach of this type of insect ten million years later. They resist stomach acid breakdown and are also very resistant to drought. Such bacterial spores may survive for hundreds of thousands of years on the surface of a planet.

E.P. Could this prove that the bacteria might even have traveled through space and survived?

K.N. Actually, survival like this has changed the way many of us think of panspermia, the theory that says germs from living beings are diffused everywhere, germs that do not develop until favorable conditions are found. We used to reject this possibility out of hand as being very unlikely. Once we saw how spore-forming organisms really can survive millions of years in very cold or very dry environments, even here on Earth, we have changed our thinking.

E.P. Ken, you will think this is science fiction, but is there any danger of bringing in a form of life that could threaten humans on Earth? Or other animals? Have you ever considered this possibility?

K.N. Yes. We always bear this in mind. We are always thinking and talking about this. The scientific community is convinced that no organism from outer space could thrive on Earth. But we cannot prove it is impossible. Therefore, we work very closely with the Centers for Disease Control in order to learn how they package such dangerous particles as human viruses when they send them around to research laboratories. We use identical methods for samples we bring in from space; not because we are afraid, but because we cannot prove that there is not any threat.

E.P. And the contrary? We contaminate the cosmos with viruses, radiation, and waste on the Moon and Mars.

K.N. In effect, and no matter how much we try, we will never be able to leave the cosmos spotless. There is a very tight control every time we send a spacecraft to the Moon or Mars. We clean and sterilize surfaces, but we know from experience, like when we go to a hospital and we acquire an infectious disease, that we can never entirely eliminate risk. This is why we expect the surface of Mars to be so difficult for bacteria from Earth to survive. If any should get there, they might survive, but they should not develop. There is no water, it's too cold, and there are no nutrients as such. We can imagine that spores might survive for some years, but they won't actually develop, like the many spores in our deep freezer. If they reached a warmer planet . . .

E.P. Like which one?

K.N. The moons of Jupiter—Europa, Callisto, and Ganymede, for example. They are very different from each other but they all have liquid water.

E.P. Are you sure?

K.N. I am almost certain that twenty or thirty kilometers below the ice of Europa there is an ocean. The problem here would be to prevent our bacteria coming from Earth to reach this ocean. Things change considerably when a planet contains water.

E.P. comments: Three years after our encounter, when Mars was being thoroughly invaded by robots (the *Beagle II* from the European Space Agency, which never landed, the *Spirit* and the *Opportunity* from NASA), Ken Nealson was among a small number of experts that never ceased to reiterate that not much could be expected from Mars, at least in the matter of finding life. "Water is not enough," Ken would repeat. "Nitrogen has not been found, and I would look for the presence of nitrogen instead of water if I were interested in finding signs of life." Nitrogen, needed for all life forms always, is involved in the energy-requiring transfer and movement of chemical components in energy transformation itself—from the inert gas (N_2) to the usable amino acids of proteins, to ammonia, and other reactive forms of nitrogen such as NH_3. Nitrogen in a reactive form is needed to make many larger molecules. There are certain universal components necessary for life. There must be some structured form of energy that allows these small components to transform from one form to another. Most importantly, these must be complex, because complexity is necessary so that all the pieces can be included and that the energy can transform and maintain matter as new and complex structure. But this complexity does not necessarily have to resemble the living beings we already know. Life from space only has to be complex, to have structure, although most likely it will have geological, biological, chemical, and physical properties different from those on Earth. This is why we search in categories such as complexity, structure, products, and energy flow, instead of insisting on the presence of protein, DNA, or other compounds already familiar to us.

E.P. Is this to what you refer when you say your team is searching for "biosignatures" in space?

K.N. I refer to properties that can be measured, that could be used to define life processes. If you read a biology book in order to know what life is, you will find that although the definition runs for over three pages, it contains nothing that can be measured. Therefore, it is futile, because we plan to send machines into space. What we want to do is to define life in measurable terms, and then design machines that can make measurements. We are talking about measurements of things that should not be there! Chemical compounds that do not fit the mold. We need a machine that will provide highly accurate information on chemical compounds, and feed this information into the computer, and every time the computer detects something that on the principles of chemistry and physics alone should not be there, it should say, "Let us take a closer look; it could be what we are looking for."

E.P. What if nothing that "shouldn't be there" is found on Mars or on the moons of Jupiter?

K.N. Don't you think it would be lovely to have a planet that never has had life? That way we could come to know the history and the appearance of the solar system independent from life. We would have what is known as a phantom member, a planet that has never had life that we can compare with our own, and contemplate the differences. And if we have one living planet and one that is not, we would be in a position to look for "bio-signatures" that are better conceptualized than those currently existing, especially beyond the solar system.

E.P. comments: In my second encounter with Ken Nealson, I saw in him the same enthusiasm that emanated from his constant obsession with life, but this time there was a slightly different perspective. Maybe he was somewhat less obsessed with water or replication as signs of life, and instead he had a greater conviction that life was a complex phenomenon. And if it is beyond water and replication, both of which are assumed to form part of a simple minimalist molecule that is a predecessor of the precursor, then it would be easier to understand that a complex phenomenon could transform the planet and the universe. This was the masterful intuition of James Lovelock, and to some extent of the founder of sociobiology, Edward O. Wilson. (And also relevant to the criticisms voiced by William Day in chapter 28.)

Are we so continuously worried about trying to govern this "spaceship Earth," this planet Earth in Buckminster Fuller's words, trying so hard to make it work, precisely because we are so busy destroying it at the same time? Is it only because here on Earth there are ideal conditions for life? Or is it that life itself has designed our planet?

Toward Perfection

INTRODUCTION

There may be more species alive today than at any moment in the past, but certainly the Paleozoic seas were replete with more diverse anatomies, far different animal body plans than persisted to the present. Perfection, like salvation, is a religious concept, one that requires fervent belief. Living beings change, none is perfect. All require continuous energy supplies to "do their thing" and for a carbon food supply to grow. All maintain their bodies to retain their form and identity by autopoiesis, that is, chemical reactions observed as incessant metabolism.

Mammals are not more perfect, more evolved, better, and "more adapted" than are insects or bacteria. With fewer than six thousand species, mammals are far less diverse than insects and their thirty million or so species. Mammals are far less diverse than bacteria, with their modes of metabolism radically different from ours. Mammals are not even the "most important" or most diverse vertebrates. Fish, with forty thousand extant species, extraordinary sensing systems and reproduction behaviors, are far more diverse than we. We are not moving toward something greater, something better, something more successful and God-like. Rather, as are other life forms, we too are changing.

The code of the dead, the DNA instructions we call genes, are present in all life forms. The genes must be preserved and passed on for any Earth life to continue. Dawkins insists that living beings are "machines" that manage their DNA, marvelous machines that appear to have been designed—but no, there is no designer. There is no purpose. There has never been any designer, nor any purpose. All animals that have ever lived were produced by Darwin's natural selection, a process observed to continue every day.

Dorion Sagan, with his colleague Eric D. Schneider, reiterates the absolute imperative of energy and material flow to any live being. And waste as gas, as liquid, and as solid must leave, must be whisked, dragged, shoved, or just seep away. In the cycling of energy, efficiency is never complete, never 100 percent, never perfect. The wastes, the spent gases of any being, still have energy or food for another form of being. The farts and burps of one type can bring clean air to another very different form of life. Sagan notes that, from a thermodynamic point of view, living beings are "open systems"; energy and matter flow through them. Energy and matter must flow. These processes, as counterintuitiuve as it seems, are necessary for the maintenance of any integrity. All systems require energy and matter flow. And all beings, in his and Schneider's opinion, "exist to reduce gradients." The world is full of gradients: all differences in energy, in matter, in gas concentration, in pressure—differences across a distance. And life's purpose, as a thermodynamic and heritable system, is to continue to reduce the immense solar gradient as the Sun bursts, releasing its heat that flows "into the cool." (See chapter 27).

Evolution and Purpose

Interview with Stephen Jay Gould

> There are fossil remains of bacteria that are over 3,500 million years old. And they are still the dominant life form on Earth.
>
> —Stephen Jay Gould

Over the past ten years, I have continued to give classes at the university, but not even in the middle of a lesson was I able to stop thinking about the ideas that the almost three hundred scientists from all over the world had explained to me or were going to tell me about the mind, life, and the universe. Literally, I have lived on a cloud, which will perhaps evaporate once this book is published.

It is very likely that I will not forget any of the scientists with whom I have conversed over this decade. I think I will always remember those I chose for

---

Stephen Jay Gould (1943–2002) was a paleontologist at Harvard University and is probably best known for his development, with Niles Eldredge, of the theory of punctuated equilibrium.

this book. The choice was completely subjective, conditioned exclusively by what I considered would be interesting for the reader, irrespective of the prestige, awards, and profundity of thought of the scientist.

But one scientist I certainly will never forget is Stephen Jay Gould—and not only for our first, and very eventful, meeting at Harvard University. We had agreed that the meeting would be at seven o'clock in the evening on a class day in his office, an immense, rather ramshackle room, with a worn-down armchair, a table without chairs, and dozens of cardboard boxes full of fossils waiting to be catalogued. It was midwinter and Stephen Jay Gould had been on an academic trip to New York City. Half an hour after the agreed time, one of his collaborators came into the office to give me the sad news that a storm had prevented the world's most famous paleontologist from catching the plane back from New York. It says much in favor of the man—Gould filled any space with his gestures and his tremendous voice—that his co-worker couldn't have been kinder and more understanding. Eager to ensure that my trip from Barcelona to Boston would not be in vain, she rescheduled the interview for the morning of the next day, between two classes, of course without consulting her boss, who was still wandering, frustrated, around a New York airport.

When, the next day, at ten o'clock in the morning, I walked down the corridor toward his office, doctors and professors were peering out to ascertain the reason for all the shouting coming from Gould's office. He was indignant at the fact that his precious time had been organized without his consent. During the first few minutes of the interview, he only answered my questions with monosyllabic yeses and nos, to demonstrate his anger.

When I edited the conversation and took out the initial series of questions with monosyllabic replies, I realized two things: first of all, that it would be the briefest of the conversations transcribed in this book; and, second, that it would be the most revealing about the contribution of scientific theory because Stephen Jay Gould had a thousand and one ways—long and short, florid and concise—of underlining the great contribution of paleontologists to modern thought. He had a plethora of phrases to make us aware that "we are the last drop in the last wave of the great cosmic ocean." The other great category of paleontologists—the Leakeys, Charles Walcott, Donald Johanson, or Miguel Crusafont—worked in quarries and caves to discover the fossil hominids that Gould placed on the map of time and complexity.

Eduardo Punset: You have often said that in the history of evolution you don't see any indication that we are advancing toward greater things, that you don't detect a line of progression in evolution.

Stephen Jay Gould: Of course. I don't know if it could exist, because bacteria have always dominated life on Earth. We are not moving toward something greater and better. And to think that we, because of the fact that we are more complex creatures, have greater probabilities of success is not convincing. Complexity does not guarantee our success in the long term. There are fossil remains of bacteria that are over 3.5 billion years old, and they are still the dominant life form on Earth. But we humans have created systems of representation of the history of life in which evolution leads to and culminates in ourselves, though the complex mammals with an insignificant number of species, fewer than five thousand, are probably not as important if we take into account the history of the Earth as a whole. For chemical and physical reasons, life must begin from very simple structures. And there is a place for complex forms of life, but the vast majority of living beings have always remained very simple, at the level of the bacteria. And, to tell the truth, they're doing very well.

E.P. Professor Gould, do you think that the appearance of the first arthropods was as important as the appearance of the first hominids four million years ago?

S.J.G. Of course, the arthropods are more important because they are by far the most widespread form of animal life. Alive today are only something over four thousand species of mammals and probably not more than forty thousand species of vertebrates, mostly fish, whereas there are millions of species of insects. Many other species of animals have not yet been identified. Without a doubt, the arthropods are, quantitatively, the dominant form of animal life today, and they will still be there when we no longer exist. We could make ourselves extinct in a nuclear holocaust and the diversity of the insects would not be seriously affected. There are over five thousand different species of beetles, for example, described in the literature. Therefore, taking into account the fact that the arthropods are in the majority and that many of the arthropods have evolved considerably, it could be said that they are the precursors of the dominant animal group on the planet. We are still in the age of the arthropods.

E.P. You note that bacteria were here before us and they will outlive us. But let's concentrate on the changes. A friend of mine at Paris University, Miroslav Radman, a specialist in genetics, has studied bacteria for over fifteen years and he says that their advantage compared to humans is that, faced with an uncertain future, they are able to mutate more rapidly. Today, there are not many advocates of change, perhaps only in certain academic circles.

S.J.G. But our change is the result of cultural change, which is much faster than any biological evolution. Biological evolution is so slow and unpredictable that it throws us into disarray. The technological advances of mankind, on the other hand, are impressive and, what's more, totally unpredictable. In the past two hundred years, we have developed ways of destroying the entire planet, of communicating instantly, of getting off the planet. . . . We can do more things than ever.

E.P. But complexity implies greater uncertainty. There are too many options and the best bet in order to survive is to bet on everything, on diversity, to guarantee it whatever it takes.

S.J.G. Yes, that is a very intelligent option, but I don't know if we are intelligent enough to take it.

E.P. And diversity still flourishes as it did in the past, doesn't it? Or was there greater diversity in the past than now?

S.J.G. There are probably more species than ever before, but in other times there were more organisms with different basic anatomies. The so-called Cambrian explosion, 541 million years ago, marked the arrival of almost all the modern groups of animals. The rare fossils of fauna with soft interiors—and not just the rigid parts—of the Burgess Shale quarry is our window on that world. It allows us to observe the true nature and the diversity of remote animal life. But diversity is not easy to achieve, and when it is lost it does not recover. If we eliminate main groups of organisms, in reality we eliminate evolutionary lines with several thousand million years of history behind them. It takes thousands of millions of years to build something, but it can be destroyed in a tiny fraction of the time needed to create it.

E.P. You have often referred to the incorrect evaluations or assumptions we humans make. What is, in your opinion, the greatest misunderstanding about human beings or the planet?

S.J.G. Broadly speaking, I think that there are two great misunderstandings. The worst we can do, and we have done it in the course of history, is to not recognize that all human beings are in fact very similar. Modern mankind is only about two hundred thousand years old. And now that we can analyze the genetic differences among human beings, we can demonstrate that, though there are differences between the different races in their external characteristics, such as the skin and the hair, it is obvious that we are very similar. There are small genetic differences among the so-called human races, but we mustn't make the mistake of succumbing to our prejudices, assuming those differences and proclaiming that our group is superior to another one. That is the sad story of racism, xenophobia, and genocide, based on a moral and biological perversion. The other great mistake we make is to believe that we are the kings of the Earth and that we have the right to decide its future. And because we have the power to do it, we don't worry about the problems we are causing to other organisms, other species, or the environment. It is a tragic mistake with potentially very dangerous consequences. We need to be more modest and recognize that all humans are one species and that we have less power than we thought; then everything would undoubtedly be better.

E.P. Should we think about time from a geological perspective?

S.J.G. Absolutely. It is the most important concept that my science contributes to humanity. The Earth is at least four billion years old, not just thousands of years. Until two hundred or three hundred years ago, in the West, it was believed that the Earth was at most four or five thousand years old and that man was the focus of the entire history of the Earth, except the first days. These ideas made us even more arrogant, as if the Earth had been made just for us and its history was the history of mankind. Now we understand that the Earth is not thousands but millions of years old and that the history of mankind is just the last fragment of a second at the end of this immense period of cosmic time. We have to understand that the Earth is not made for us, that we are simply guests who are here thanks to a lucky accident. Perhaps this idea will increase our respect and our humanity.

E.P. From a geological perspective of time, are there any indications of new massive extinctions?

S.J.G. We know of five massive extinctions in the 550 million years of the existence of animal life on Earth. In other words, an extinction occurs approximately every one hundred million years—not very often. Of course, there has never been another species like humans on Earth, with a unique consciousness, for better or worse, and with so much power. Many species have become extinct because we have changed their habitats, and so many species are disappearing that it is possible that mankind is causing a massive extinction. We should be more intelligent, to prevent that from happening. The future will tell, but a massive extinction has never meant the complete disappearance of life. Life goes on and we are here. If the dinosaurs had not died out sixty-five million years ago, we would not be here now, because the dinosaurs would have dominated all the small mammals for hundreds of thousands of years, until another important, external phenomenon appeared. The fact is, the disappearance of the dinosaurs made it possible for the little mammals to evolve. And that is why we are here.

The Code of the Dead

Interview with Richard Dawkins

> Some animals live in the trees, others in the sea, others underground, others fly, others dig, but, essentially, they are all doing the same thing: they are working to survive and therefore to pass their genes on to the future. The genes encode the instructions that enable animals to exist.
>
> —Richard Dawkins

Richard Dawkins and Stephen Jay Gould are the great scientific popularizers of the twentieth century and, I dare say, the twenty-first century, too. Although they have greatly advanced science and, at the same time, have brought it closer to public understanding, it is very probable that Dawkins will never receive the

---

Richard Dawkins holds the Charles Simonyi Chair for the Public Understanding of Science at Oxford University.

Nobel Prize. The Nobel Prizes are only given for the discoveries of scientists who call themselves "drillers," rather than visionary scientists infused with overwhelming intellectual curiosity. Both Richard Dawkins and Stephen Jay Gould belong to the second category.

My first meeting with Richard Dawkins took place in Munich, at a seminar organized by John Brockman. Almost two thousand young people packed the room.

To Dawkins we owe the idea, formulated in books like *The Blind Watchmaker* and *The Selfish Gene,* that in our form and in our content, we are short-haul machines or taxis that transport genes in their never-ending race for survival. But his vision of evolution and, particularly, of culture, goes much further. He maintains that the twenty-first century is digital and that the digital discontinuity has made modern electronic technology viable. In a way it is related to physics and biology.

"In order to live, we use a code of the dead: the genes of our ancestors," Richard Dawkins, popularizer of Darwinian ideas, said to me.

Richard Dawkins: Nerve fibers don't just work with analog codes; rather, theoretical calculations demonstrate that they would not work without digital codes. Nerve impulses are like shots coming from a machine gun. The difference between a strong message and a weak message does not lie in the intensity of the impulses but rather in their frequency. Seeing the color yellow, hearing the sound of the sea, smelling the smell of turpentine—the differences between these sensations are processed in our nervous system by means of the evaluation of the shots or impulses. If we could listen to the inside of our brains, it would sound like a battlefield.

Eduardo Punset: What consequences will the economic, social, and personal convulsion caused by the appearance of nanotechnology have?

R.D. Nanotechnology consists of creating tiny machines, at molecular scale, and it has a very promising future. Given that molecular biology is pure natural nanotechnology, in a way we know what can be done. Molecules like enzymes and proteins work by catalyzing specific chemical reactions. These molecules are like tiny mechanical instruments, tools at molecular scale, and there are millions of them. We will be able to design machines of molecular size, in principle, it seems, but they are only effective if millions of machines are used at the same time. Therefore, they have to be duplicated—and that is viable in the case of molecular structures. We can design a single one and then duplicate it by means of a production process. Then we would release millions of machines inside the organism, which would repair breakdowns, cure certain diseases such as cancer, and would act in certain treatments in which the surgical scalpel is too aggressive.

E.P. We work like machines.

R.D. All life is a series of marvelous machines. Each living being is a machine capable of managing its DNA, and each animal does it in a different way. Some animals live in the trees, others in the sea, others underground, others fly, others dig, but, essentially, they all do the same thing: They are working to survive and, therefore, to pass their genes on to the future. The genes encode the instructions that enable animals to exist. This is a wonderful part of universal thought; in each case, the details of the instructions are different. A mole is prepared to dig the earth and eat worms, while the genes of an eagle instruct it how to fly and hunt its prey. But, in all cases, the key is the survival of the instructions themselves. I reaffirm this concept.

E.P. You have been much criticized for maintaining that genes are more important than people in the evolutionary process.

R.D. Individuals are very important because they are the ones that live. They walk, observe, hunt, escape, group together. But we have to look through the individual to see what is happening on the inside. In fact, both levels are important. We cannot ignore the individual, but if we want to truly understand it we have to look at genes, the fundamental instructions of the DNA, which permit the development and behavior of the individual.

E.P. Richard, do you think that the genes are tired or getting old? Or are they being destroyed? Or do they remain as exuberant as ever?

R.D. We should take this idea into account. The evolution of life has developed over the course of thousands of millions of years, and for hundreds of millions of years, more or less, there have been the same forms of life, the animal. It might seem that genes are getting tired and that things should change with the passage of hundreds of thousands of years, but no, they are as exuberant as ever.

E.P. And isn't genetic information lost in the process? They say that some chromosomes, in their process of division, lose fragments in which there was information.

R.D. Yes, that happens all the time, but it is part of the general process of mutation: fragments of chromosomes are lost and other fragments differentially multiply. The genome, with the passage of generations, is not fixed. It does not constantly grow. Rather it is a flow of information. The genome has always been mutating and always will. The general organization has been the same for three hundred million years.

E.P. When we look at the beautiful things of this world—a wonderful bird, a crystalline river, if there are any left—the question arises whether it has been designed by someone or is the result of a process of natural selection. In other words, has the world been created, or resulted from natural selection?

R.D. There is no design whatsoever, on the Earth or in the universe. The best candidate for "design" has always been life. Life has always survived, until it has become, with human beings, the most perfect phenomenon that can ever be designed. Living beings are so beautiful, so elegant, so complicated—and every fragment of a living being looks as if it has been designed. The eye seems to be designed to focus on an image and analyze it, hearing to hear, hands to pick up objects, legs to walk. But this is not true. Nothing has been designed. Darwin demonstrated the devastating, simple, and true explanation of how everything happened in 1859. The result of Darwin's mechanisms, the survival of characteristics and natural selection, is the development of incredibly complicated machines that seem to have been designed. The sensation of design is almost perfect, although certain imperfections are detected in the design of life forms. Certain errors are interesting because they would not exist if life had been designed. They are the result of the influence of history and natural selection.

E.P. I would like to quote a couple of expressions of yours about genes. You wrote that genes are "a codified description of a distant past," "a survival manual," and "the genetic book of the dead." I think that's extraordinary.

R.D. It is a Darwinian consequence. The genes are like a survival manual of a past environment, because it was in the past that these genes overcame natural selection. Today, we know how to survive, and any other animal can survive if the past codified in its genes is not very different from the present. Animals manage to survive because their ancestors survived. When the present diverges greatly from the past, then their lineage is split apart because they are adapted, "designed" for the past and can't survive. The consequence is that present genes are a description of the past, of a world in which our ancestors survived. Thus my phrase "the genetic book of the dead."

E.P. And today's extremely rapid changes, like technology, could they be a genetic threat to the manuals of the past?

R.D. Yes, every radical change in the environment is a threat. The ice ages, droughts, the periods of great earthquakes, comets, and meteorites from outer space are a threat. The meteorites, in particular, exterminated the dinosaurs. Human technology and civilization are a threat because they have radically changed the world. Because of human influence, the world has changed drastically compared to what it was like in the past. And almost all the animals, except those that can evolve very rapidly, like the mosquitoes, the flies, and the bacteria, are adapted to live in a prehuman world. The human world is hostile to them. Some species, like rats and seagulls, adapt very well; they change their way of life.

E.P. Do they also adapt to catastrophic changes?

R.D. Yes, they take advantage of the changes humans have caused, but other species cannot adapt; that is why so many are becoming extinct. Though this doesn't mean humans deliberately cause their extinction.

E.P. And what will happen to humans? Will we be able to survive?

R.D. Things are going surprisingly well for humans, on the whole. It's true that there are many mental pathologies and alienations due to the fact that some humans feel they are in a strange environment, but in general, as a biologist I am impressed that our species is able to tolerate life in this anti-natural environment.

E.P. It could be argued that, as well as the information codified in the genes, there is now also information that has been learnt and stored in the brain that gives rise to technology. And, for the first time in history, the two channels of knowledge are colliding, aren't they?

R.D. Yes, the idea fascinates me. My expression "the genetic book of the dead" comes, obviously, from cerebral knowledge. It's like turning the same analogy round. You are right, we have two separate lines of knowledge from the past that are, in turn, knowledge on how to survive, that is, the knowledge of how the species has survived in the recent past. And one of these lines is progressing much faster than the other one, because cultural changes are so much faster than genetic ones. Genetic changes have not stopped, but they are so slow that—

E.P. We hardly notice them.

R.D. Or we do not notice them at all. To understand human history, we would have to forget the whole theory of evolution and start from zero. The animal that lived in Africa in the Stone Age has not changed much, but it is growing under the influence of cultural knowledge. Therefore, the most recent book of the dead, that of culture, is growing, though I don't know if genetic and cultural knowledge will come into conflict. Perhaps the only foreseeable catastrophes are that cultural progress will not be able to cope with catastrophe, or that cultural progress produces such advanced technology that it is dangerous. Today, we have weapons that can destroy the world in the same way that the meteorite wiped out the dinosaurs sixty-five million years ago! Living with these bombs is one of the conditions of our future.

— 27 —

Purpose of Evolution

Interview with Dorion Sagan

> The purpose of life to break down a gradient is an epochal development on a par with the great Copernican realization that we are not at the center of the universe.
>
> —Dorion Sagan

It seems that heat spreads into the cool. It never goes back from the room and warms the coffee cup. Dorion says that this is our clue to the nature of life's behavior and stock market behavior. What does he mean when he says, "Life breaks down a gradient"—that abhorring and breaking down gradients is obeying a thermodynamic law?

Dorion Sagan is a prolific science writer.

Eduardo Punset: Dorion, in your book with Eric D. Schneider you claim that the "purpose of life" is to break down a gradient, where a gradient is a "difference across a distance."

Dorion Sagan: Thank you, Eduardo. I consider the idea that the purpose of life is to break down a gradient to be an epochal development on a par with the great Copernican realization that we are not at the center of the universe. A gradient is indeed a difference across a distance. The essence of the second law of thermodynamics, which has long been thought to be simply an "increase in disorder," is not. We know this in part because the second law applies throughout the cosmos, which is full of strikingly organized systems, such as life itself. The second law is more accurately—and simply—characterized by saying that energy has a strong tendency to spread. Energy delocalizes; it goes from being concentrated in one place to being dispersed. This, not the increase of entropy construed simply as disorder, is the essence of the second law. The work of Frank Lambert, integrated into virtually all recent chemistry textbooks, makes clear that the second law is really a matter of energy dispersal, not increase of disorder. There is good new evidence that the function of life, as of other naturally occurring complex systems, is to produce entropy, or atomic chaos. These systems maintain their designlike shape and cycle matter as they disperse energy. That is their natural function, or purpose.

E.P. I think you mean that the solar gradient, the sun's radiation and its heat, goes, as you say, "into the cool" and life breaks down this gradient. But maybe, do you mean the geothermal gradient? We know that some life, like the deep-sea vent animals, live without light, and there the gradient might be more the hot interior of the Earth relative to the cold ocean at the surface.

D.S. Life, which is one of these systems that cycle matter in regions of energy flow, uses the sun. But other systems use other gradients—chemical, pressure, and temperature differences—from which energy can be derived. The other systems include things like Bénard cells, Taylor vortices, Liesegang rings, Belousov-Zhabotinski reactions, salt fingers in the ocean, Hadley cells in the atmosphere, and storm systems such as tornados and hurricanes. All of these systems produce entropy as heat. They are organized and have a natural function, to reduce a gradient. The hexagonal Bénard and other convection cells reduce a simple temperature gradient. Tornados come into being to destroy the difference between high and low air pressure masses. A tornado appears, comes spinning into existence in a highly unlikely configuration of cycling atoms that persists until its natural job, using the energy potential of the gradient, is done. After that it ceases to exist. Its purpose, the reason for its existence, is to reduce the barometric pressure gradient.

E.P. There is something lifelike about it.

D.S. Yes, but there is nothing mystical about it and no design, in the sense of a humanlike creator overseeing the nonrandom constitution that is the powerful and purposeful tornado. The tornado's form and function are one: to spread energy and destroy gradients.

E.P. And life?

D.S. Life is a similar gradient-reducing system, only it measurably reduces not the atmospheric pressure but the electromagnetic radiation difference between the hot sun and cold space. Satellite measurements of Earth's biosphere show that, because of high albedo, reflective cloud cover over rainforests in the tropics, lush ecosystems at the equator are losing as much heat in the summer as does Siberia in the winter. As emphasized by atmospheric chemist James Lovelock, who was hired by NASA to look for life on Mars but found our planetary atmosphere displays a continuous thermodynamic anomaly, life likes it cool. So do other complex thermodynamic systems, which produce entropy as heat into their surroundings. The second law's inanimate purpose of delocalizing energy is beautifully accomplished by open-cycling thermodynamic systems such as tornados and ecosystems. More atomic chaos—entropy or "disorder"—is produced by these systems than would be without them. These systems disperse energy but they are not disordered. Far from it. Highly organized, they actively tap into energy to promote entropy production. Not only does life not violate the second law, it allows the second law do to its thing more effectively—either more rapidly, or over a longer period of time, or both.

E.P. But how do the deep sea communities break down the solar gradient, then? Don't they use geothermal energy?

D.S. No. Life does exist in lush quantities at the bottom of the ocean in regions of heat vents and "black smokers" exuding geothermal energy. There are even a growing number of scientists who believe that the sort of iron-sulfur chemistry occurring without life under such conditions was the direct chemical ancestor to life's origins. So the gradient-reducing system of life may have come from the gradient-reducing system of nonlife. But the base of these possibly primitive underwater ecosystems is not geothermal energy. Heat is a thermodynamic end product, not a source of energy for life. The vent ecosystems are supported by bacteria that gain energy from the natural oxidation of hydrogen sulfide. In this case a chemical gradient, the so-called redox gradient, is tapped into for energy. This may have been life's original gradient, and then it migrated, after photosynthesis evolved, to the solar gradient. Today, as geopolitical events suggest, global technological society depends upon another gradient—the fossil fuel gradient. The animals you mention, the giant pogonophoran tubeworms you can see in James Cameron's wonderful Imax film, *Aliens of the Deep*, may like it hot but they do not live off a geothermal gradient. They are fed by their internal bacteria that gain energy from the sulfide gradient.

E.P. Please explain what you mean by "the purpose of life" and its connection to thermodynamics. I thought thermodynamics was a physical science, but you imply a relation to biology. Can you please tell me what you mean?

D.S. You are right. Thermodynamics is a physical science. But now we see that life is a physical phenomenon, not only in terms of its chemical makeup as genes and proteins but also in terms of its energetic process as matter cycling in regions of energy flow. The "purpose of life" is of course a loaded phrase. The great problem with scientific treatments of teleology, the study of purpose, is that the default position of all questions of purpose is that they are unscientific. There is a good reason for this. The pre-Renaissance worldview saw the entire cosmos as a purposeful place in which humans participated. The scientific revolution created a sharp partition between purposeful humans and inanimate, mechanically working nature. Science made great accomplishments by being vigilant about attributing the purposefulness of the human mind to gods, spirits, or forces outside subjective consciousness in nature. That said, the trajectory of modern science has been an unending triumph of realizations that we are not special but part of a lawful material universe of surprising beauty if not whim. Copernicus catapulted us out of the center of the universe but his ideas were scoffed at as unscientific at first. Bruno was burnt at the stake and Galileo held under house arrest for being arrogant enough to adhere to the nonarrogant position that we were not the geographical center of creation. Then the chemists found that life's substance wasn't special. Urea, an organic chemical, could be synthesized from inorganic constituents. The astrophysicists found carbon, oxygen, hydrogen, formic acid, and alcohol in the depths of space. In the quip of Tallulah Bankhead, the Alabama socialite, daughter of a senator, and witty partier, "I am as pure as driven slush." We too are not pure or special but creations of our cosmic context. If we neither live in a special place, nor consist of uncommon elements, so too is the process in which we are involved not radically unique. Thermodynamics shows us that we have the same entropy-producing function as other naturally occurring matter-cycling systems that grow in regions of available energy. Although life has been identified as, and reduced to, its genetic architecture, its capacity to chemically copy itself, this ability is ultimately secondary to its energetic function. The function of life, like the function of a tornado, is to reduce a gradient. This does not rule out some sort of higher metaphysical purpose, but it does present us with the startling realization that a credible materialistic reason for life's physical existence has finally been found.

E.P. And what about evolution? Does that have a purpose?

D.S. Just as thermodynamics reveals cycling, energy-using life has a function, we can now see that evolution has a direction. It is true that evolutionists such as Stephen Jay Gould [see chapter 25] and Richard Dawkins [see chapter 26], who disagree with each other on adaptation and genetic determinism, agree that evolution is essentially random. However, there is a growing body of evidence to show that this orthodox position is wrong. Over time life has spread in area from the oceans to the land and the air and now, increasingly, space. The number of chemical elements and the depth of their involvement in the biological circulation at Earth's surface has increased. Since, given an energy source, all thermodynamic systems grow insofar as they are able, cycling matter as they do so, this should not surprise us. Other evolutionary trends exist. The number of species, despite five major mass extinctions, has bounced back each time to achieve a higher number. The cumulative energy stored in the biosphere and routinely deployed by life has also augmented. The respiration efficiency, the amount of energy derived from oxygen, has increased. Arguably, the net intelligence, perceptivity, and responsiveness of life—all of which can be used to find and deploy new energy sources, as well as protect the means of access to them—have also increased over geological time. Where before we might have been tempted to deny the evidence for this direction in life because it suggested some sort of mystical entelechy, we can now see that these trends of evolution are natural. They parallel the activity of nonliving thermodynamic systems. When we look for food for dinner we are supporting ourselves as gradient-reducing systems. When we seek dates and mates we are increasing the chances that our type of gradient-reducing organization will persist in the next generation—despite the body's inevitable decay and death, itself a form of entropic dissolution. There is no longer a need to bridle at talk of life's purpose. Even nonliving systems have a function, a purpose or function, from which life's purposefulness is derived. At the same time, the purposefulness of life is not that of the grand anthropomorphically oriented cosmos of prescientific times. This purposefulness is natural, Copernican, and shared with other, nonliving systems.

Dead or Alive?

INTRODUCTION

Life, all forms of it, is composed of cells. Nothing less complex than a watery cell, a unit composed of hundreds of proteins and bounded by an active membrane that takes substances in and lets them out, is alive. Indeed, a cell, whether a bacterial or a nucleated cell (the only two choices), as the minimal living system, shows all the properties of life simultaneously. Life requires incessant flux of both expendable energy and chemicals. Life's chemical needs—carbon, nitrogen, hydrogen, and so forth, are most familiar to us as food. The energy, which must be supplied in an appropriate and usable form, goes to maintain the system in its active state; when all the energy is used up the cell dies. A few kinds of cells (or organisms) can dry out, shut down, and wait: spores, cysts, some seeds, active dry yeast, and other propagules. The chemicals known as food need to be steadily supplied, as well; they are incorporated into the cell. Any cell or organism made of cells form systems, and these systems tend to disintegrate with time. All living systems are always comprised of a long list of parts. Maintenance takes continuous energy and food substance (nutrient) flow. Life is "running to stay where you are" to paraphrase the words of *Alice's Adventures in Wonderland*'s White Queen. The properties of life shown simultaneously by live cells include movement at least at the molecular level, sensitivity to various aspects of the environment (e.g., mechanical and chemical stimulation), and the tendency to grow and reproduce. Since movement at all levels ceases at temperatures of absolute 0°K (–273°C), the properties of life in some cases can be preserved at this subfrozen extreme. (The frozen life-form, the cold-tolerant propagule, will show all its normal properties simultaneously when rewarmed.) Life itself in these cases is retained by the cell throughout the ordeal of freezing and thawing.

All of the scientists I know who have participated in the "origins of life" field have concerned themselves with the chemical activity. Following the fabulous lead of Stanley Miller and Harold Urey (1953) they have attempted the generation and formation of complex carbon-hydrogen (a.k.a. "organic") compounds from simpler gas mixtures. Or they have tried to make genetically significant compounds (nucleotide or nucleoside components of DNA or RNA) in water or in "organic solvents" (formic acid, alcohol, acetone, cyanide, etc.) by all sorts of tricks. The literature abounds with the assumption William Day and I call the "cake mixture approach." Bring together the relevant ingredients in their proper proportions under appropriate conditions, treat them with a suitable energy flux such as ultraviolet radiation, cosmic rays, or heat, and voilà: life will jump out. Nearly all of the "origins-of-life scientists" have been chemists. Although Bill Day, too, is an experienced chemist, he has abandoned any "cake mixture approach." He claims, appropriately I think, that life is and has been "a growth process" since its inception at least 3,500 million years ago.

I feel Bill Day's message should be more widely available, and my personal hero, the expert biologist-chemist Professor H. J. Morowitz (George Mason University) agrees. So I exercised my co-editor's prerogative and included his interview. Eduardo approves. Other interviews in this section on real life are authentic Punset innovations. One, that of my closest colleague, Professor Guerrero, nicely shows the uninitiated what are the living elementary particles of the biosphere. The bacteria in the broad sense invented life and have run the Earth since the earliest Archean eon. The astounding capabilities of bacteria, indeed, their wisdom, is virtually unknown to those who have not studied the field of "microbiology." Microbiology is not traditionally "covered" in academic biology courses. Guerrero's discussion opens us to a vast world, huge diverse populations of which we are dimly familiar, as a culture, with only a tiny number of freaks such as the anthrax bacillus and the food-poisoning salmonella.

If Guerrero shows us prokaryotes (chapter 29), J. T. Bonner (chapter 30), the renowned and profound biologist, introduces us to one of my favorite groups of eukaryotes: the amoeba and its multicellular descendant, the slime mold he has studied for so long. Zoologists who belong to the ratliver–*E. coli* axis are a dime a dozen. Botanists who live and work with plants have written many splendid books. Mycologists, like Bryce Kendrick (2003), reveal the whole world of fungi. Therefore, to represent life we complete our readers' acquaintance with "real life" by our introduction to the prokaryotic bacteria and the eukaryotic amoebae, our own microbial ancestors who lived for 3,000 million years before any animal evolved on Earth!

— 28 —

Life as Growth

Lynn Margulis's interview with William Day

We must entirely abandon our preconceptions in this field.
—WILLIAM DAY

Bill Day, alone and with a few students and colleagues from the University of Mississippi, has done laboratory experiments on relevant gases, mineral solutions, and small organic compounds to test his central idea that life is not a Stanley Miller–Harold C. Urey–Leslie Orgel "cake mixture." Never in life's long history has life been such "a thing." Life for him is a growth process. Day has ideas about how to start simply to produce the energy flow and chemical interactions that might initiate a growth system comparable to the one that he infers must have existed at the origins of life on Earth. He has had some

---

William Day conducts independent research on questions of structural order and the origins of life.

success in his laboratory experiments. When he was invited to Spain to describe his work in 2004, however, he declined the invitation. He was not ready to present his scientific results to nonscientists. He is not yet sure of enough of them. Hence he missed his potential interview with Eduardo Punset and participation in our "science as culture" program at Conde Duque Media Lab in Madrid. Luis Rico, director of Media Lab, had asked me (LM) to invite Day to participate in science-cultural activities associated with his *Banquete* (*Banquet*) show. Rico's program took place inside a huge tent in the form of a giant book set up with 250 collapsible chairs. What else? It was a replica of volume two of *Don Quixote* by Miguel de Cervantes, the literary figure in the Spanish language comparable to Shakespeare. It is appropriate enough that the four-hundred-year anniversary celebration of the publication of *Don Quixote*, by the way, a book written in prison, should include science—especially the science of the "origin of life." Anyway, had Bill Day been less reticent to come to Spain to discuss his scientific "work–in–progress" and had Eduardo been able to interview him, it is clear to me that the interview would have gone like mine here. Delighted to have the opportunity to present Day's ideas, the "growth-concept" that mainstream "prebiotic chemists" have tended to reject, usually prior to publication, I interviewed him by telephone and e-mail based on his written descriptions. I used Day's papers in good faith to write this version for this book, and received his grateful approval for the inclusion of this chapter. I hope that next year Bill will go to Barcelona or Madrid to enjoy a nonvirtual *Redes* interview with Eduardo.

Lynn Margulis: I think you disagree with many scientists who follow the "prebiotic chemistry" point of view, those who claim that if we can make genes (nucleic acids, I suppose) under conditions close to those inferred to have been widespread three or four billion years ago, then we would have a reproducing life system and would be on our way to producing "life in the laboratory." Or at least we would have a better idea of how life itself came about so long ago. What are the implications of your work for the problem of the origin of life?

William Day: Many "origins-of-life experiments" have produced organic compounds of significant complexity from gas mixtures taken to be representative of the Archean eon atmosphere. After all, amino acids were generated in the laboratory from sparked methane, ammonia, and water vapor. The famed Miller-Urey experiments in 1953 and their successors produced formaldehye and simple sugars, hydrogen cyanide and its polymers, including even adenosine triphosphate (ATP). They made orotic acid and other pyrimidine precursors and lipids with ease. However, since these achievements fifty years ago, I maintain, we have seen very little progress in these sorts of "prebiotic chemistry" experiments. The resolution of this "origins-of-life problem" ought not to be more difficult than any other fundamental scientific inquiry. However, my claim is that we do not understand what life is. Since we don't know what it is we cannot properly explore and explain its origin! We have no clear idea of what it would mean to solve the "origin-of-life-problem" in the laboratory or in the field. This lack of a scientific principle to define life has impeded interpretation of many accomplishments in the literature. It has failed to inspire new relevant experiments.

The ways in which all living matter differs from all chemistry needs to be recognized. Until now the scientific investigators assume that life's properties result from the precise composition and organization of its components. They inferred that for life to originate prebiotic substances just assembled into structures of sufficient complexity and then began to function.

I disagree entirely with this assumption. I call it the old "cake mixture" concept. I do not think that if one brings together in proper quantities life's essential ingredients—amino acids, lipids, nucleotides, et cetera—and just mixes them with an adequate energy source that life will ever emerge.

L.M. Why won't this work?

W.D. No single cell, no living body, is ever static; living beings dynamically produce themselves by regeneration. The living system itself breaks down its components by lytic enzymes and resynthesizes anew its parts, as Schoenheimer and co-workers showed in the 1930s already.

L.M. What did they show?

W.D. Well, the universal turnover is never chemical equilibrium. Rather they showed the animal body is a steady state of metabolic reactions that continually consume energy. Any animal deprived of appropriate energy flow, any that is impeded from the continuous metabolic breakdown and resynthesis of its parts, dies. In fact, any form of life acts this way. When the energy and material flow that fuels buildup and breakdown is disrupted for whatever reason, synthesis processes cease. Growth stops. The system fails. The unbalanced degradation reactions continue until thermodynamically unstable structures form. Life's system collapses. Life is irreversibly lost. There is no going back.

L.M. So, what does this imply? I think so many of life's chemical processes are reversible.

W.D. Yes, they are, but only as single processes. The system is not symmetric. We must recognize life as always a system directional in time. Any given system, given organism, does not reverse. Its spent energy or food components are not reusable. Life is a set of incessant processes; to know this clarifies approaches to the deciphering of its origin. Life's chemical processes are unique, the system is all-inclusive. By its nature, the interlocked chemistry of any life-form is dynamic. In the flow of material and energy that define the process of its structure, composition and dynamics are always inseparable! Life is less like a clock or a cake that can be assembled from its components and then started up or "turned on" than it is like a waterfall, a dynamic, ever-changing system that forms as it proceeds. Synthesis, autocatalysis, and reproduction are simply aspects of one and the same biological process. I suggest that, until now, in the great body of literature the authors assume that they study the origin of life. These guys like Stanley Miller and Gerald Joyce, Leslie Orgel, and David Deamer don't study "selves." Rather than study the origin of life, they have made good progress in the study of the "origin of food."

Researchers tend to separate the features of the living: the limiting membrane, the genetic-protein synthetic pathways, autocatalysis, and metabolism as if they were steps relevent to the origin of cells. Their mechanistic approach is completely flawed. Rather the cell's constituents, synthesized by the system itself, must be continuously removed by transformation, dissolution, or some means. Life began with the minimal dynamic process itself and all the features of the cell were integrated into all preceding features as the cell evolved. Life from the beginning controlled its chemical reactants. The autocatalytic cycles, membrane, metabolic sequences, and genetic system developed in sequential and hierarchical order. We need to understand that sequence, that hierarchical order or we will be stuck in a rut for another fifty years!

L.M. I guess you think we have not made progress in the origin-of-life problem at all?

W.D. That's right. No real progress since the late sixties or early seventies.

L.M. Is the situation hopeless?

W.D. Not at all! As soon as we realize that life as a process has never been lost since it began, we will recognize the utterly essential and universal aspects of the core process in all life. We have clues: iron-sulfur clusters that transfer electrons; pyrophosphate kinases that connect high-energy phosphate transformations to carbon-dioxide incorporation into organic matter; and glutamate that comes from the conversion of ammonia from the environment to glutamic acid in cells. These are the cores of ancestral, continual energy flow and they were with life since its inception.

L.M. We need to change our perspective?

W.D. Yes, we must entirely abandon our preconceptions in this field!

— 29 —

Our Ancestor, the Bacterium

Interview with Ricardo Guerrero

> We humans are only a sparkle in the eye of the evolution of a persistent, sturdy, powerful, single huge life-form on planet Earth.
>
> —RICARDO GUERRERO

Ricardo and I have worked together on *Redes* (*Networks*, my TV program) for years. Whenever I need authentic, reliable information about any aspect of the entire of microbial world, I call him. If he does not know he finds someone who does. This "interview" was patched together from some dozen interviews I've done with him, mainly in Barcelona.

Ricardo Guerrero is Professor of Microbiology at the University of Barcelona and also the editor-in-chief of the professional scientific journal International Microbiology.

Eduardo Punset: Why do you think bacteria are so important?

Ricardo Guerrero: All the microbes are important, not just to me but to all of us. Especially the bacteria. I'd like to tell you about their unseen power. They are the ones that chemically transform the environment and formed the earliest ecosystems.

E.P. Wait a minute. What is the difference between "bacteria" and "microbes?" Aren't all microbes bacteria?

R.Gu. Not really. All bacteria are microbes. They are small organisms that are made of cells that lack nuclei—even if they are multicellular. "Microbes" are simply organisms, life forms—most easily seen with a microscope. So—though all bacteria are microbes—there are other kinds of microbes also, for instance, yeast. Yeast, all single-celled fungi, are eukaryotes; their cells have nuclei. There are many different kinds that evolved separately from distant fungal ancestors. They are microbes but not bacteria. And so are all the smallest algae (like diatoms and *Chlamydomonas*). So are the amoebae, the paramecia, and the dinoflagellates microbes. But yeast, diatoms, and other microscopic algae, amoebae, *Paramecium* (the common pond-water ciliate), and *Plasmodium*, the causative agent of malaria, are all microbes. But they are nucleated microbes: eukaryotes.

So we have two great groups of microbes: bacteria without nuclei and all the rest we see under the microscope, the nucleated or "eukaryotic microbes."

The bacterial group is called "prokaryotes." And all the nucleatal ones are "eukaryotes." They (algae, ciliates, trypanosomes, giardias, amoebae, etc.) have nuclei. Each cell contains at least one nucleus . All the nucleated ones are either yeasts (or other small fungi) or protists; none are bacteria or archaea. Bacteria, prokaryotic microbes, were the first inhabitants of Earth, which is 4.55 billion years old. Maybe the origin of life took place on our planet as early as 3.85 billion years ago or even earlier. While it is possible that the initial features of our two neighbors in the solar system, Venus and Mars, were conducive to the appearance of life, most likely, life could not have been maintained on either planet. Life started bacterial. But something special happened on Earth that allowed life to persist: the establishment of ecosystems among prokaryotes. Prokaryotes, bacteria, were the first. They were the basis on which all other forms of life evolved. Over the first 2 billion years of bacterial evolution, almost all metabolic strategies developed. Symbiotic associations among prokaryotes gave rise to the ancestors of all the complex and varied life forms that followed and now exist on Earth.

They also served as the origin of the eukaryotic cell, in the form of protists, which were the first eukaryotic-celled organisms. Indeed, all such nucleated organisms—i.e., protists, fungi, plants, and animals—emerged within a prokaryotic world. The eukaryotes have retained intimate connections with, and dependency upon, prokaryotes ever since. For example, the mitochondria, the organelles inside all our own eukaryotic cells that are responsible for oxygen respiration, come from a proteobacterium. Similarly, chloroplasts, the photosynthetic bodies in plants and algae, descended from a group of aerobic photosynthetic bacteria called the cyanobacteria, the blue-green bacteria that used to be called "blue-green algae."

E.P. You believe that the earliest ecosystems, at least I think you believe they were also bacterial, no?

R.Gu. Yes. Bacteria inhabit all possible locations in which life exists, from those offering "ideal" conditions for growth (from the point of view of us "macroorganisms," of course) to those representing extreme environments—habitats too hot, too salty, too high pressured, too acid et cetera, for us animals and plants.

Microbes, especially bacteria, are responsible for cycling the elements for life that include carbon, nitrogen, sulfur, hydrogen, phosphorus, and oxygen. By cycling these elements in soils, microbes regulate the availability of plant nutrients, thereby governing soil fertility and enabling the efficient plant growth that sustains human and other animal life. Bacteria also play an essential role in recharging atmospheric gases. Bacteria are master recycling experts, too: they can degrade biological wastes and release products for use by other organisms. Only recently we have become aware that microbes are the bases for the functioning of the biosphere. Thus, we are at a unique time when the interaction of technological advances and the exponential knowledge of the present microbial diversity begin to allow significant advances not only in microbiology—but also in the rest of biology and other sciences as well.

E.P. So, you claim that the earliest ecosystems, I believe in the Archean eon more than 3 billion years ago, were bacterial ecosystems. Please explain to us, most of whom are not biologists or ecologists, what exactly is an ecosystem? Give us an example of a bacterial ecosystem and then a nonbacterial ecosystem.

R.Gu. An ecosystem is a minimal volume on the Earth of life with its coupled environment, a volume in which all the chemical elements needed for life are entirely cycled and recycled inside that ecosystem. A microbial mat is an example of a bacterial ecosystem usually dominated by the photosynthetic cyanobacteria. A forest or a coral reef or a grasslands would give us examples of nonbacterial ecosystems. The bacteria are still there but the producers are trees, coral animals with symbiotic algae, or grass plants in the "macroorganism" ecosystems.

E.P. When were the earliest microbial ecosystems?

R.Gu. What made the maintenance of life on Earth possible was the development of these microbial ecosystems. They, the earliest ecosystems, prevented all the elements on the planet's surface needed for life from being used up. This squandering of elements would have taken place in a maximum of 200 or 300 million years after life originated. Without early ecosystems early life would have been extinguished. Since then, the establishment of ecosystems where the products of the metabolism of some microbes served as food for others allowed the recycling of food matter. The activity of these ecosystems determined the planet's subsequent evolution. Indeed, until approximately 1.8 billion years ago the whole planet had bacteria, only prokaryotes lived here. Bacteria were its sole inhabitants. Life is a geological force; it acts on the planet in a way that maintains the necessary environmental conditions for life itself. Adaptation to temperature, chemical composition of soils, and the proportion of gases in the atmosphere was and is necessary for life to continue. Eugene Odum (1913–2002), the ecologist from the University of Georgia in the United States, supported the idea of the biosphere as a homeostatic system composed of living beings. Life might be an ineluctable outcome of planetary evolution, a natural continuation of the physical and chemical development of the universe. We know there is life on Earth—some say that it is even intelligent—and we know there is no life on the Moon. The twenty-first century will tell if there is or was life on Mars, or Venus, or in other parts of our solar system such as on Europa (Jupiter's moon) or Titan (Saturn's moon).

E.P. I know you and your colleagues think life, microscopic life, is everywhere. Others are afraid of microbes. To them microbes are "just germs." How can we make visible the "invisible" to ordinary people?

R.Gu. Microbes are everywhere! This ubiquity of microbes is based on three things: their small size for easy dispersal, their metabolic—what I mean is their chemical versatility, their amazing capabilities, and their genetic plasticity. They pass on and acquire genes to and from each other. This allows bacteria to tolerate and adapt quickly to unfavorable and/or changing environmental conditions. In many bacterial ecosystems, the functionally active unit is not a single kind or population of bacteria but a consortium of two or more types of bacterial cells that live in close symbiotic association; they live together.

Despite the fact that microbes are everywhere, they were only discovered very late in the history of humankind. Antony van Leeuwenhoek (1632–1723), a middle-class draper, a man from Delft in the Netherlands and without any scientific training, observed the first microbes (protists) in water in about 1674. He called them *beesjes* (beasties), or *kleijne Schepsels* (little creatures), as he wrote in old Dutch. Or *animalculi* in Latin, as it appeared in the published translation of Leeuwenhoek's letters to the Royal Society in London. Later, in 1683, he observed bacteria, smaller than his protist "beasties" on the surface of his own teeth.

E.P. After Leeuwenhoek how did the "invisible" become "visible?"

R.Gu. As I see it, the discovery of microbes involves at least three stages: the microscopic stage, the pathogenic stage, and the ecological stage. For since Leeuwenhoek it was known that microbes existed and that they had a defined form in space and time. The microbes at first were considered mere "curiosities" for the human intellect. No one thought that those minute creatures had any function. Then, with the "fathers of microbiology," Louis Pasteur (1822–95) and Robert Koch (1843–1910), people became certain that microbes caused infectious disease and contaminated food and water. Now, in the third stage, thanks to the pioneering investigations of the Dutchman Martinus Beijerinck (1851–1931) and the Russian Sergei Winogradsky (1856–1952), microbial ecology started. We have learned that the activities of microbes completely control the cycle of the recycling matter in all ecosystems, and from the beginning have controlled the evolution of the biosphere.

E.P. You confuse me. First you say microbes control the environment's conditions, then you say they cause disease, contaminate food and water. How can they be so good, so important, and so bad at the same time?

R.Gu. How can a wonderful student, first place in his mathematics class who always helps his mother and does his work on time become a murderer of Iraqi aunts, mothers, and infants? Didn't he just "become bad" when given a rifle, an assignment he couldn't avoid, and when he was sent with the rest of the army? Only then did he become like the rest of them, no? His conditions changed. His own survival was at stake. The same with microbes. We humans live in a dynamic state of coexistence with myriad microbial life forms. Some are usually mutually beneficial as they keep us healthy, others under other conditions can be adverse. We humans still are here because the microbes that attack us have an interest in our survival. Our deaths are accidents of collateral damage. We survive because it is not to any microbe's advantage to eliminate us. Instead, the aim of microbes that devour us, what drives their evolution, their natural selection, is not to eat us up and finish us off but to domesticate us. Prolonged interaction between humans and our infectious microbes, carried out across many generations and among numerous populations on both sides, creates a pattern of mutual accommodation that allows both man and microbe to survive. A disease bacterium that kills the victim quickly creates a crisis for itself. It must somehow find new victims quickly enough to keep its own chain of generations going. A human body that resists bacterial infection so completely that the would-be microbe who eats us cannot find any food or lodging creates another kind of crisis of survival for the bacterium. The most successful symbiotic microbes are those who persist in healthy or near-healthy victims and promote and exploit their behavior for two principal goals: to ensure continuity of healthy forms, both the microbe and the person, and to promote efficient dissemination from one to another human.

E.P. So not only does the same microbe behave differently with different conditions, but microbes themselves are very varied? Is this what you are saying?

R.Gu. Yes. The same population of microbes changes. We always deal with populations, growing populations, not individual bacteria.

E.P. How are microbes, bacteria, different from each other? Especially those that interact with people?

R.Gu. The great diversity among microbes, diversity of structure, of mode of reproduction, of ecological relationship with us and other species, and evolutionary history makes the subject of human interaction with them so challenging. Groups, populations of different bacteria, fungi, and protists, do cause most infectious diseases. The incredible diversity among such microbes results in varied modes by which they meet us. *Staphylococcus aureus* that is involved with skin and other infections, *Mycobacterium tuberculosis*, the causative agent of tuberculosis, and *Helicobacter pylori* give us three examples of disease bacteria that have very different living infection strategies within us.

E.P. For example?

R.Gu. We have only time for one example—the most recently figured out.

The 2005 Nobel Prize in Physiology or Medicine was awarded to the discovery of *Helicobacter pylori*, its role in ulcers, in the field of bacteriology. In fact the discovery was made in 1982 when J. Robin Warren and Barry J. Marshall in Australia first described the presence of *Helicobacter pylori* in the human stomach and related it to subsequent gastritis and peptic ulcers. Marshall, right here in Barcelona recently, presented his Nobel lecture with the title "*Helicobacter*: The Good, the Bad and the Ugly," a direct reference to one of the most famous of Sergio Leone's films.

Warren and Marshall showed that the presence of *H. pylori* in the stomach was always associated with inflammation of the underlying gastric mucosa. Their recognition that *H. pylori* caused ulcers meant to the doctors that they had to make a major paradigm shift. Ulcers could no longer be considered "psychosomatic"; instead ulcers became an "infectious disease." Physicians had noted the presence of these spiral bacteria in the human stomach as early as 1906. Although similar observations were repeatedly reported after that, they were ignored because the bacteria could not be grown outside the stomach in the lab. The normal human stomach was touted to be sterile because it was considered "too acid." But bacteria taken from the gastric mucosa were grown in the lab, outside the stomach, by 1982. Marshall and Warren also described spiral or curved bacilli in tissues. The spiral bacteria were often seen in stomach cancer or in ulcerated gastric tissue. But the idea that growing bacterial populations, not stress or too much acid, were the cause of ulcers contradicted all medical dogma! Then came trials that showed conclusively that elimination of this helical bacterium dramatically improved the clinical course of ulcer. But Barry Marshall himself did it: he swallowed a broth, an active culture growing in liquid, millions of *H. pylori* bacterial cells, and soon thereafter personally developed gastritis. He had horrible stomachaches, the prelude to ulcers. But he cured himself from the disease after treatment with antibiotics! Warren, Marshall's senior, did not join Barry in this dramatic experiment because he already suffered from peptic ulcer. Then the two successfully treated other people suffering from ulcers. They clearly identified the *Helicobacter pylori* microbe as the culprit. *H. pylori*, in 1994, was the first bacterium to be classified as a "class I carcinogen" according to the World Health Organization criteria.

Now *Helicobacter pylori* has been found in the very acid stomach contents of people all over the world. In so-called developing countries, 70 percent to 90 percent of the population carries *H. pylori* in the stomach, while in "developed" countries infection rates are lower, from 25 percent to 50 percent. The colonization and persistence in stomachs for many years shows that *H. pylori* is a common stomach bacterium. We can see now that *H. pylori* must be confronted with multiple environmental challenges. How does it enter? Adhere to stomach tissue? Resist gastric clearance? Protect itself against the corrosive effects of stomach acid, or the immune system? How is it transmitted to other humans with different genetic backgrounds or stomach conditions?

In recent years, other *Helicobacter* strains have been discovered in different species of mammals, both in unhealthy and healthy animals. The detailed nature of the symbiotic relation between the helical bacterium and the animal stomach is under study.

E.P. Ricardo, that is a fascinating example of human-bacterial relations with different outcomes, but what about eukaryotes? I thought that infectious "germs" are bacteria treatable by antibiotics. Eukaryotic microbes are like us, are in fact composed of the same cells as we are, so how could we treat diseases caused by eukaryotes with medicines that don't make us sick or kill us. Shouldn't they kill other eukaryotes? Do protists, who like us are eukaryotes, do they ever cause disease?

R.Gu. Yes, they do. Most infections, but certainly not all, especially in warm tropical countries, are bacteria. So let us consider another scourge, a protist scourge, of humankind: malaria. Some historians say the beginnings of malaria date as far back as ten thousand years ago, even earlier. The establishment of agriculture and farming created favorable conditions for spreading the disease. The increase in human population density that followed the shift from a nomadic to a sedentary agricultural way of life that included cattle may have provided breeding places for anopheline mosquitoes. Archaeological deposits in the eastern Mediterranean show malaria was there at least six thousand years ago. Ancient products used to treat it were remarkably effective. They confirm the antiquity of the disease. A drink made from qinghao (the plant *Artemisia annua*) has been used for at least two thousand years in China. The active chemical ingredient, artemisin, only recently has been identified.

Reference to malaria is found in early Chinese, Chaldean, and Hindu writings. Since the fourth century B.C., malaria has been endemic in the Mediterranean. Fevers caused by the proximity of humans to swamps and standing water were noted by both Greeks and Romans, who drained swamps to control periodic fever episodes. Despite the antiquity and morbidity of malaria, it is unclear whether the disease was ever as serious as it has been in modern times. During World War I, thousands of soldiers on the Macedonian front were victims of malaria attacks. But, curiously, this was not the case in the battle of Actium, in September of the year 31 B.C., when the Romans (under the leadership of Octavian, who later became Emperor Augustus) confronted the Alexandrian army led by Mark Anthony and Cleopatra. The battle involved some four hundred thousand soldiers, but apparently no losses were due to malaria. Why? The disease already existed. What was the reason? By the fifth century A.D. apparently the malaria mosquito carried the protist Plasmodium that changed from a mild variety to another, much more aggressive pathogen. Can we speculate that, had the later, very efficient malaria protist lived in Europe during antiquity classical culture with its influence on modern civilization worldwide would not have developed? We don't know.

E.P. OK, Ricardo, you have convinced me of the importance of microbes. I just have one last question: you tell us that we all evolved from bacteria. But in what sense are bacteria our ancestors? I thought primate mammals were our ancestors.

R.Gu. Eduardo, you are right! Bacteria are our oldest ancestors, primate mammals are not only our most recent ancestors—they are us! The differences between bonobos, the ordinary chimp, and us are trivial. And we have many more ancestors as well: the earliest bacteria whose cells were composed of DNA, RNA, proteins, and smaller molecules wrapped by membranes. These formed swimming, fermenting, tightly organized bacterial consortia, communities that evolved into our nucleated cell ancestors. These formed multicellular units that became our soft-bodied invertebrate animal ancestors, some of whom evolved into fish, and then later some reptiles who evolved into mammals who eventually gave rise to our mammalian, including our primate, ancestors. The outlines of evolution of the single, immense "life-form" with its many parts are clear. We humans are one very recent blink, only a sparkle in the eye of evolution of a persistent, sturdy, powerful, single huge life-form on planet Earth. But my message is that all life ultimately is bacterial and all evolved from bacteria. Furthermore, all other eukaryotic living beings on Earth still today depend on prokaryotic, bacterial life. Prokaryotes are present in all the places where any life can exist. They occupy the widest variety of environmental conditions that range from the "ideal" growth conditions (ideal, obviously, from our "macroorganism" point of view), to extreme environments (unthinkable to animals, plants, or other recent forms). Without knowledge of microbes, biology would be much more limited. We would not know that life thrives in conditions of extreme temperature, salinity, or pH. Photosynthesis would only be a metabolism of aerobic, oxygenic plants and algae. But no, it was "invented" by bacteria, by prokaryotes, both anaerobic and anoxygenic!

E.P. I now see why you are right: The Earth is a bacterial planet. Life originated in the form of bacteria and developed as a phenomenon of bacterial communities of all kinds and in all places. I have to change my negative attitude to the subvisible microbial world. Since it represents the most powerful life force on Earth, my ancestors, my technological allies, and probably even the most numerous and successful form of my descendants, it is time that I learned to respect, to even love, the hardy denizens of the microbial world! Thank you, Ricardo, for helping expand the life forms for which I now feel a greater empathy than ever before.

— 30 —

Different from Amoebae

Interview with John Bonner

> What is important is the ability to communicate, and not how communication is carried out. This is true for all organisms.
>
> —John Bonner

It is surprising to note—even for me, not that I am hesitant—how many times I have made a wrong decision with regard to fundamental issues throughout my life. During the fifties when I came back from the United States, where I discovered how a particular system worked, I decided to join the Communist Party after I read for the first time the *Communist Manifesto*, obviously in the United States and not in Spain, where it was banned for decades. It was foolish not to have anticipated the fall of a sectarian system, such as the Soviet one, that was based on the suppression of individual rights.

John Bonner is the George Moffett Professor of Biology Emeritus at Princeton University.

The night of February 23, 1981, when Lieutenent Colonel Tejero and his troops tried to overthrow the government in a military coup during a session of the Congress of Deputy Leaders, I was Minister of Relations with European Communities in Spain. I had just returned from twenty-five years abroad. After several hours of silence imposed by the Civil Guard, one of my colleagues asked me what went through my head when they started shooting into the air and shouted, "Everyone on the floor!" I answered, "Why on earth did I choose to return at a time like now?" This is just an example of—at least in my case—the poor capacity of the brain to handle information.

I used to be convinced, a conviction that I shared with scientist John (Joan) Oró, that the first satellite photographs of planet Earth would convince everyone that the borders that separated countries were not real. It seemed ludicrous, after the oneness of the Earth was seen from space obviously lacking in any hint of national borders, that people would not question the existence of the superimposed legal or political borders. Real national borders do not, in fact, exist. Unfortunately, overwhelming evidence that there are no borders is not convincing. Today nationalism—ethnic, cultural, and political—is more alive than ever.

Another example of my capacity for self-deception, as clear as the others mentioned but more relevant to Bonner's interview here on animal culture, relates to genome sequencing (see chapter 19). Time and time again I have said that when society realizes that our DNA is identical to that of the other animals, from the *Drosophila melanogaster* fly to chimpanzees, repentance will sweep through society for the despicable way in which we treat them. No one will kill cats at night, throw stones at dogs, or drop goats from steeples, bleed bulls to death, or invasively experiment on chimpanzee brains in laboratories, because no one could commit such heinious crimes against one's own close relatives. Even though humans are tetrapods, related to their coelocanth and lungfish ancestors, our most recent marine relatives, once again I was wrong.

From recent revelations by cognitive scientists, my poor ability to predict is less due to any innate unique traits of mine, than that all humans actually know much less than we assert when it comes to predictions of the future. Research by many scientists shows our mediocre modern hominid attempt to forsee our future.

One thing is the decipherment of the human genome, and quite another is "conscience" or "higher intelligence" behavior that, apparently, is inherent in humans. The four reknowned scientists who study animal culture in this book are Jane Goodall of Gombe (chapter 3), Jordi Sabater Pi of the Barcelona Zoo (chapter 4), Nicholas Mackintosh of Cambridge University (chapter 1), and John Bonner of Princeton University (chapter 30). They all reject the often-unstated assumption of the enormous chasm, the huge and unequivocal differences, believed to exist between humans and other animals. They all can see that humans and other life-forms including other animals share features characteristic of "higher intelligence."

The French paleontologist Yves Coppens explained the reasons behind the long-standing nonsense that goes on within the beautiful walls of l'Académie des Sciences de Paris. Two groups of paleontologists differ quite markedly. The first includes the great academics, who are well versed, fundamentally, in public understanding of science through ideas more than through fossils. Paleontologist fossil-knowers have brought an understanding of geological time to humanity. They provide a new dimension to modern work in the same way that the painters of the fifteenth century discovered the secrets of perspective. The vision of our place in the universe changed dramatically owing to the geological view of deep time. Paleontologists such as Stephen Jay Gould, and his work on the natural assembly of mid-Cambrian animals preserved in the Burgess Shale quarry, have re-created the first steps taken by our direct ancestors about 500 million years ago (chapter 25).

Yves Coppens, however, belongs to a second group, those who dig for fossils. Paleontologists who have largely spent their lives with their hands in the soil like Coppens are credited (together with Maurice Taïeb and North American colleague Donald Johanson) with the discovery of the first complete skeleton, called Lucy. This first hominoid *(Australopithecus afarensis)* dates back more than two million years. Fragile and petite with a cranial capacity only one-third of that of her descendants, Lucy was already a person from head to toe.

"Invariably," says Coppens in his slow style, "the same sequence is always repeated." In the excavations we first find a biological change produced by unknown mutations, and shortly after we find the technical change that gave rise to the biological one: a supposed improvement, such as the shape of tools. But then hundreds or thousands of years go by before the cultural change that brought about the biological change actually appears. Early cultural change is nearly as slow as genetic change. The frustration that this delay produces makes us question: Is cultural change the result of new genetic variation or are technical advances inevitable? What actually happens is that the geological time perspective is innate in cultural change. The social and political organization patterns, or the creation of a new paradigm of scientific knowledge, follow a much slower process than does technical or institutional change. The discovery of the structure of the DNA molecule will eventually change our relationship with the rest of the animal kingdom even though we still do not know when or how.

John Bonner, professor emeritus of evolutionary biology at Princeton University, has immense knowledge about the similarities of humans with other animals and other, even stranger life-forms. His studies center on social amoebae (the slime mold protoctists here). These are one of the earliest kinds of multicellular nucleated organisms from which we have much to learn.

Eduardo Punset: You have been giving lectures in this room for the last forty years. Have you ever explained to your students the differences between human and social-amoebae intelligence—between us and the tiny social moldlike creatures that you have studied all these years?

John Bonner: To tell you the truth, I have never explained this difference in class. But I now wish my students were here to listen to me, as I have asked myself the same question time and again. I am interested in the intelligence of all animals [slime molds are not animals], from amoebae to humans, and I have worked with social amoebae since I began biology. People assume that because they are mere amoebae they are stupid. In fact they can be very smart. First of all, amoebae live in groups rather than independently. They form sacks of amoebae with front ends and rear ends. Within a given space, amoebae can migrate a few centimeters. It is extraordinary. They move toward a source of light with incredible precision. They are capable of moving toward a very weak source of light. Just a tiny amount of light is enough to attract them. Years ago we discovered that they move along heat gradients with incredibly small differences. They are highly sensitive to heat. Therefore, I have the impression that they do act with intelligence, but only as a group. They are incapable of these behaviors as single amoebae, but when a hundred thousand or even a million amoebae join together to form something more or less a millimeter in length, they are very smart indeed and respond to environmental stimuli.

E.P. Incredible! Together they are much more intelligent than as individuals?

J.B. Yes indeed. It is very interesting how they do this. In the case of light, for example, amoebae are translucent so that light is actually propagated through them, but because they are cylindrical in shape they form a lens where light is concentrated at the rear end and thus creates a gradient. To compensate for this gradient, the front end, which receives the light but transmits it toward the rear, moves faster. They move toward the light, in other words, but only because the light is concentrated at the rear when they form a sack. The movement in search of light is produced because they exercise this social capacity. No single amoeba could do this, but as a group they can.

E.P. Do you think the same rule applies to humans? Do you think we are more intelligent as a collective than as individuals?

J.B. I think so, even though it is not very scientific to make such simple assertions. The point is that large brains work better than small ones, which is to say that in certain ways there are analogies. Many people say that I work with slime because I can manipulate it to my heart's desire and it really does not care. In fact the slime molds almost seem to like it.

E.P. Really?

J.B. We cannot really know if a ball of amoebae has fun, but it does seem that way to me.

E.P. Don't they die? What I mean is if they have the ability to regenerate.

J.B. Without a doubt. If we split in two one of these cylinder amoebae that I refered to before (they measure approximately a millimeter in length) each half will recover, move, and produce normal slime quickly.

E.P. In this, social amoebae are more intelligent than we. They say that one of the greatest comparative advantages of humans over other animals was, at a particular moment in evolution, the ability to guess what another person was thinking. Therefore, guessing what someone else is thinking involves predicting, in part, their reaction, thereby preparing for self-defense, or to help the other or somehow manipulate the situation. Do you think it is true? Does it apply to animals?

J.B. I would say this is true and that it works the same way in other animals. I think that two dogs—and I have owned dogs—know perfectly well by the way each acts, what the other is thinking. They can predict behavior. Both human and animal intelligence should always be regarded in a social context, to my understanding. Human intelligence is not isolated. Rather it results from communication and dialogue between people, and I believe this is exactly the same for other animals.

E.P. How do you define intelligent behavior? When you are able to deceive or to communicate with someone? How can we determine if a behavior is a subproduct of intelligence?

J.B. Before I answer you, I want to say a few things. Firstly, I am not a great believer in definitions, because they impose strict limits. You cannot enclose everything within the boundaries of a definition. I believe, and this follows the previous point that all intelligence is a continuum, some animals show very limited intelligence, and progressively speaking, brains become more and more complex until you get those that have adapted to think and perform complex skills, like ours. I prefer the idea of continuity and admit the difficulty in defining intelligence.

E.P. Therefore you believe that intelligence is a question of degrees and not, as many hold, that we are intelligent and other animals are not.

J.B. Exactly. I apply the same argument to consciousness. An old friend of mine, the biologist Donald Griffin, has put a twist on the big question of whether only human beings are conscious by asking, "How do we know that animals are not conscious of their actions?" He provided several examples that make one question if indeed animals have some sort of self-awareness, although different from our own—

E.P. Self-awareness and the ability to think about how others act—

J.B. Other animals have an understanding of their behavior. They are aware that they have behavior. Don't ask me as you did before regarding intelligence, what is consciousness, because that is even harder to define, but I am convinced that there is a continuum of consciousness among animals.

E.P. Therefore I cannot ask you what the difference is in culture—an even more ambiguous word—between other animals and humans. Is culture also a continuum?

J.B. Yes, I define culture as a continuum, but not so that it applies exclusively to humans, as do many anthropologists. If we define culture as the transmission of information from one individual to another by way of behavior or dialogue, we deal with such a very simple definition that it applies equally to the lower animals and to human beings. But this definition infuriates cultural anthropologists because they believe that the term *culture* belongs to them and that biologists like me have no right to question it. They define culture in terms of civilization and other concepts that refer exclusively to humans.

E.P. You talk about models in nature, forms that have certain predictability or order, and have written so many fascinating books and articles related to strength. You almost suggest that physical constrictions determine the size of the animal, its culture or behavior.

J.B. You have used the precise word, constriction. Size, for example, is very important because in life body density is that, roughly of water—a little more than 1 gram per milliliter. Thus animal weight increases as does its cube. Size is related to weight. But strength and surface (like skin) are functions of the area, thus squared, not cubed. If you compare the legs of an elephant with those of a gazelle or elk, the elephant legs are very thick, as Galileo reasoned. A large animal such as an elephant requires such thick legs, otherwise it would not be able to hold up its body. These are not strengths that impose limits. Think of embryology and development: the embryo of an elephant foresees how large the adult will be and starts to develop thick legs. This is not determined by the model, but rather the reverse. The animal will grow large and requires a development plan to ensure it will be well proportioned. Nevertheless, what you said at the start is true; the elephant cannot reject the constrictions of its large size but the constrictions are not the main deciding factors.

E.P. I would like to ask you another question that intrigues me. Is it correct to say that all animals have, more or less, the same cerebral potential, and the only difference is that they employ this potential in different ways? A dolphin in water uses sound waves to communicate with other dolphins because sound waves travel easily in water, for example. A kangaroo, a lion, or a human employs the same potential of light waves, because they work better on land. A flying bat uses echolocation, sonar. Biologists, for example, Jordi Sabater Pi, believe that other animals are not actually less intelligent than people but rather they focus their intelligence on fields that leave us at a loss [see chapter 4].

J.B. Absolutely. Your examples refer to means of communication. A dolphin emits sounds like we do, but dolphin sounds are especially effective for travel through water. I have been fascinated always by the discovery that elephants emit such low-frequency sounds they are undetectable by humans. These are cases of communication, they do not really reflect intelligence. Without the ability for communication and expression, it is true that no animal could be intelligent. Intelligence lies in the ability to communicate, not so much how it is done.

E.P. You refer to sociability of amoebae, no?

J.B. Exactly. Dolphins have a very complex social structure, while—

E.P. And a rich family life—

J.B. Yes. Male dolphins swim in groups of three in order to more effectively subjugate a female as a result of teamwork. Their social life is extraordinarily complex.

E.P. May I delve a little deeper into an idea you have supported throughout your career as a renowned biologist? An organism is neither adult nor child, but rather an entire life cycle, and therefore not properly reflected in a static photograph. What do you mean exactly?

J.B. I explain how I came to that idea. At university, I studied a course on logic, even though I did not like it. In one of the first classes the professor indicated that when we use a word or the name of a person we are reminded of the person at a precise point in his life, but it is all fictional because the embryo is already that person. Therefore it is inaccurate to isolate a single period of time, even though to do so is practical. Each animal is a life history and not the instantaneous "photograph" that we observe in the films or the blinds and observation towers.

E.P. Or a fantasy of the imagination.

J.B. The life of a dog, or of any other individual animal, spans from the time it was an ovum to the time of death.

E.P. What is the difference between conceiving life as an instantaneous photograph at a given moment in time and imagining the complete life cycle from a practical perspective?

J.B. I think the big difference is that if we think of a human being as a life cycle we are mindful of the development: its growth from earliest first cellular phase—where cells divide and form the vertebrae and spinal cord—through infancy, immature adolescence, and culmination as a mature person. I believe we should see humans as wholes.

E.P. It is surprising that human beings stop physical growth at a certain time in their life cycle. Yet fish, turtles, and many other species of animal continue to grow until they die. Is this a curse or an advantage? Why the difference?

J.B. I cannot say whether it is good or bad, but certainly mammals differ greatly. Even elephants continue to grow. It is not enough to say mammals; the difference includes humans. The difference you mention is related to the way growth occurs. In our twenties our long bones are sealed; we stop growing taller, unless we have pituitary abnormalities. This is related to Darwin's natural selection. There are reasons why humans and their ancestors, like many mammals, understood that growth cessation was advantageous. There was an attempt to conserve ideal height to facilitate movement to escape from predators or to hunt, I believe. This is related to the limitations of size. But why do elephants continue to grow? Perhaps because when they are so big and heavy already whether or not they are a little bigger or smaller does not really matter.

E.P. Let us return to the issue of intelligence and differences, if they exist, between other animals and people. Animals do not cultivate art or aesthetic thought, right? Where did artistic thought come from?

J.B. This has always fascinated me. During the twentieth century we uncovered the sense of beauty in animal life, as though it had been strictly human. But in the nineteenth century, the Victorians praised the beauty of the plumage of an animal, its colors, because they admired the animal's artistic sense. Gradually, these ideas became the object of mockery and they became very unpopular. During the twentieth century, male birds-of-paradise were said to possess extremely beautiful and elaborate plumage employed in courting females. Another conspicuous but not the most attractive of birds, the bowerbird, built a very elaborate structure he decorated with fruits of the forest and colorful shells to mate in.

E.P. Like a nuptial chamber—

J.B. Exactly. "Bower," in fact, is a Victorian term for a chamber of love. They believe that when two species split in the course of evolution, one group of birds preserved their bright colors, which has certain disadvantages (brightly colored male birds make easier prey) as opposed to the bowerbird, which does not run that risk. Only their nuptial chamber, and not the bird itself, is gaudily adorned with color. To return to our discussion, I believe strongly in continuity and that birds possess a sense of art. Perhaps it differs from our own sense, but possibly humans have preserved that part of the bird's brain that is sensitive to art. I am fascinated by the fact that gorillas and chimpanzees draw if you provide them with a piece of paper and a pencil or brush. Anyway, we have not answered the question if animals have a sense of art. Nevertheless, I think that birds and mammals do things that, in their own way, look somewhat like our concept of art.

E.P. Does art originate for humans that way, as well? In birds sexual selection channeled their artistic abilities. . . .

J.B. This is a very important question. In many tribes, the ornamentation of the body, i.e., tattoos, scars, foot binding, was tied to sexual selection. Even though obviously these practices are not genetic, they do have an intelligent purpose. It is very surprising that precisely during the Cro-Magnon period they painted wonderful rock and cave paintings that had nothing to do, it seemed, with sexual selection. Rather they may have expressed cosmic religiousness. To be honest, primitive man possessed great artistic capacity.

PART FOUR

TOWARD THE INVISIBLE

From the Vast to the Miniscule

INTRODUCTION

Science studies the visible universe, all of it, and in many cases must infer the immensity of the invisible, dark-matter universe. The objects of scientific study extend far beyond the limited skin of a woman or the madness of a football crowd. Scientists attempt to understand the very smallest particles of matter and the very beginning of all life as well as life's possible futures. As my brother-in-law Sheldon Lee Glashow (chapter 33) explores, we, all people, are middle-sized. We are far too large to contemplate atoms on their own level and far too small to have any perspective on the vast universe, visible or not. The interviewees here tell us space is mostly empty, so we walk through a cloud of electrons with no direct experience or sensation. If "consciousness" means "awareness of the environment around one," one dictionary definition, then atoms, too, must be conscious.

The invention of small and smaller machines to do our work for us is all the rage these days. What do they mean, anyway, by "nanotechnology"? *Nano* in Latin (*ennano* in Italian) means "dwarf," or "very small." We stand on a ladder. All of us are used to locating ourselves in a single place, a dot on a three-dimensional sphere north (of the equator) and west (of the Greenwich UK meridian) and one meter above the ground. But Einstein told us this cozy place on the ladder is only relative to fixed points on static Earth's surface. He showed that we needed for our full localizability another dimension. We need to add the fourth dimension: time. So here I am 40° north, 65° west, one meter above the ground on my ladder, and it is January 1, 2007.

I was entirely happy with four dimensions, our family x, y, z, and time. Indeed I was grounded and oriented until Lisa Randall, with passion, told Eduardo that we have mentioned and paid attention to only the four ordinary dimensions, whereas the existence of many more is extremely likely. At that point in this narrative the final interviewee, Paul Davies, boldly asserts that we all have restricted imaginations. Both we scientists and you readers limit our own horizons. Davies has a way to reorient us all in new time and space coordinates. What we thought we knew and understood in 1900 about the scientific world, a world of orderly measurable "facts" wrangled from tangled banks and starry skies, is all wrong. Now in 2007 the scientific world, even as recent as 1999, is receding to be replaced with a new Hubble Space Telescope and Mars Rover images. Worms, pogonophorans from deep-sea vents that lack organs but are full of packed populations of bacteria, and astounding underwater coral reefs that entirely lack photosynthesis have been seen off the coasts of more than forty different countries, beginning with Norway. Science is the only news but even its news is always changing. Keep reading. Keep learning. Keep challenging received wisdom. Ask any investigator, political leader, and TV personality to justify his or her opinion with evidence. Seek the crucial observations. In honest inquiry some things remain. Doubt of authority, insistence on criticism from within, need for detailed observation, and emphasis on the knowable—these aspects of science do not change. As a social, cooperative, and metanational activity this mode of learning in science does stay the same. But the mind, life, and universe, the narrative unveiled by scientific inquiry continues to astonish. These revelations, "facts" in constant need of verification, come to us only with difficulty. Indeed, science is the only news.

— 31 —

Walk through a Cloud of Electrons

Interview with Eugene Chudnovsky

The electrons in the ground repel the electrons in our shoes
and that is why we walk on a cloud of electrons.
—EUGENE CHUDNOVSKY

Lamarckism fit the Stalinist regime like a glove. According to the French biologist Jean Baptiste de Lamarck, acquired knowledge, as well as character traits and transformations developed or imposed in the course of an individual's lifetime, were transmitted to the following generations. The transformation of the world and the creation of the new Communist man would be ineradicable. It was possible to avoid our dependency on parsimonious and random genetic mutations. Of, if you like, genetic instructions were sensitive

---

Eugene M. Chudnovsky is Distinguished Professor in the Department of Physics and Astronomy at the Herbert H. Lehman College in Bronx, New York.

to the generational imprints imposed by the Communist regime. Exactly the opposite to what was—and still is—the scientific consensus in the Western world.*

For many years, the standard bearer of the official doctrine in the Soviet Union was Trofim Denisovich Lysenko, the president of the all-powerful Academy of Sciences of the Union of Soviet Socialist Republics. Lysenko's influence began to wane at a famous Congress of the Academy, which Eugene Chudnovsky also attended. Lysenko said, in defense of the official theses, "If we cut the ears off calves when they were born, generation after generation, after some time cows would be born without ears." "Professor Lysenko," timidly asked a young scientist whose face was lost somewhere at the back of the room, "if it were true that by systematically cutting off cow's ears, generation after generation, they would end up being born without ears, how do you explain that all the young women of the Soviet Union continue to be born virgins?"

Eugene Chudnovsky loves to remember the uproarious laughter that immediately broke out in the room, the beginning of the end of Lysenko's prestige and, with it, the official dogma. As a good scientist, Eugene questions established thought and speculates on what follows; he enjoys inquiry. So I asked him, "Is our perception of the world correct? I mean, does it reflect what the world is really like?"

Eugene Chudnovsky: I think it reflects the world as it is, and what's more, with great precision. Our brain, which is what analyzes the world, has been developed over millions of years of biological evolution to reflect the world and model the processes of this world so that we can better adapt. Therefore, if it did not reflect the outside world correctly we wouldn't be able to model it, nor function adequately.

Eduardo Punset: There would be mistakes. Something like that must happen with people who suffer from a mental disorder.

E.C. Indeed. If we take the example of people who abuse drugs, they will end up having a different perception of the world and will not be able to function correctly.

E.P. Do you see the universe exactly as I see it? The stiff, leafless tree through your office window—

*Although Lysenko's version of Lamarckism is invalid, the idea that all genetic novelty is generated via random mutations is an invalid, misconceived overgeneralization, as well; the role of symbiogenesis, a Russian concept more than one hundred years old, as a major source of evolutionary innovation is indispensable. (See chapter 21 and Margulis & Sagan, 2002.)—LM

E.C. I think so, given that we are people of comparable intelligence. What we really see is an image that depends, mainly, on our brain's ability to process a certain amount of information. For example, if we compare the processor I have inside my head with that of a falcon, a bird flying in the sky, the falcon may have a much more perfect system of vision than mine, but it is incapable of keeping the image in its brain with the same details as I can.

E.P. Nor for as long.

E.C. Exactly, nor for as long simultaneously. Therefore, the falcon's image is, probably, though I don't yet know how to demonstrate it, like a large gray screen with a mouse moving in the middle.

E.P. comments: I have never forgotten the image suggested by Eugene, of the falcon seeing a gray universe crossed by a single mouse: a kind of large, faded blanket like the Plaza de Catalunya in Barcelona with a large rat—also gray—crossing it from one side to the other. And I have often thought that the day on which science brought down the wall that supposedly separates us from all other animals, our perception of the world might be different, but it is equally self-interested. We see the universe that is appropriate for us, in accordance with what our brain thinks we need in order to survive. It is the commonsense strategy to which Richard Gregory refers in chapter 15 in this book.

It would seem logical to assume that everything we know, we know through the mind: either what is observed enters the mind, or the mind affects our perception of the world. One of the most beautiful solutions to the continuing debate between reductionists and antireductionists I heard from the biologist Harold J. Morowitz is the idea that we see what we need to see in order to survive.

Starting from the mind, we postulate the floor, the chairs, the table; we construct objects starting from ideas. And to explain what a chair is in reality, we can structure the theory of atoms and molecules; and to explain what molecules are in reality, we can structure the theory of electrons and protons. And, finally, we can go right down to quarks and string theory. That is the way of the reductionists. But we can also do the opposite—that is, we start from the particles and we ask ourselves: how is it possible for atoms to create living substances? And how is it possible for living substances to create more complex living organisms? When, in this way, we come to the animals, we see that they have developed another property, which is cognition. So, instead of being solely the result of random combinations of atoms, they can carry out actions with a purpose. And all of this brings us back to the mind. It is like a circle.

Reductively, we come to the fundamental particles, though we never reach the true basics; and, on the other hand, we recreate the mind through an understanding of the things that formed the origin of life and evolution—until we reach the mind.

Morowitz suggests a conciliatory, elegant way of closing, transitorily, the debate between reductionists and antireductionists. By definition, the definitive resolution of the mystery of mysteries can only come when we know for certain not only when the inflationary universe began—which we do know—but also how and why.

E.P. Do you think we can maintain the pretension of knowing or of modifying what will come after us in the evolutionary expansion? I mean: do you think we have any influence over what will happen after us?

E.C. I think we have a very limited influence. As humans, we only decide the future of technological evolution; we can even prevent that evolution from going in a harmful direction, though to a certain extent we are incapable of avoiding—for example, terrorist acts—even if we try. However, I don't think we can prevent the appearance of new species as a result of the technological revolution; these species are comparable to biological species that have evolved. The new species I am talking about are computers, which, without a doubt, in the near future, will process information faster than we, and will acquire abilities.

E.P. But do you believe in cyborgs, in those half-machine, half-organism hybrids?

E.C. Of course I believe in cyborgs.

E.P. And in the possibility of them reproducing?

E.C. I don't know whether or not cyborgs represent the future of evolution. This is impossible to predict. Of one thing I am sure: we are not well built, I mean physically, to face the next steps in the evolution of life, such as space travel and exploration of planets. We need a specific environment, temperature, pressure, and chemical substances in order to survive. Our skin is incredibly vulnerable. We are also very exposed to pollution, because the majority of our bodies are made from elements in the first half of the periodic table. And, what's more, with technology we are contaminating the environment because technological goods are made, above all, with elements from the second half of the periodic table, which are heavier.

E.P. You mentioned temperature. . . .

E.C. Temperature is important because we can only live in a very narrow range, unlike the robots of the future, which will have almost none of these limitations; they will be able to go into space without protective suits, reproduce on planets or in spaceships, and conquer the universe.

E.P. When you talk about space, I think of the emptiness around us, even inside us. Our molecules are made of atoms, and around the nuclei of those atoms electrons revolve at considerable distances compared to their size. They are sustained in the air. What state of affairs is that? I mean, are atoms stable, or rather are the electrons that orbit around the atoms stable? Who's to say that they will always remain there?

E.C. Yes, the electrons will remain there. Unlike the satellites that revolve around the Earth, because the physics of satellites is classic physics: due to friction—some air always remains in the orbits in which they move—in time, in the end, they will fall. But the electrons will not fall onto their nuclei because their movement is different and is explained by quantum physics, which is not at all easy to understand. It is true that the electrons, the nuclei, and the atoms exist in empty space. If we're talking about a gas, like the air in this room, the distance between the molecules is about ten times the size of an atom. So there is much more empty space in a gas than in liquid or solid matter.

If you go into space, the distance between particles like electrons might be one meter. Therefore, most of space is empty. In any case, I make no reference to what we call dark matter, because we don't know where it is. Perhaps dark matter is everywhere, occupying all of space. Perhaps it is not composed of particles. Today, physicists believe that matter is not composed only of particles. In the past, say, twenty years ago, it was said that everything was composed of point particles, a kind of very small, we could say spherical, particle. But now we know that there are objects that are not composed of atoms and they can be very extensive objects like strings. . . .

E.P. Strings?

E.C. Yes, "strings." These are solutions to mathematical equations that derive from the so-called theory of everything. These strings can be very large objects, comparable to the size of a star or a galaxy. But they are not made of atoms.

E.P. Are they vibrations?

E.C. Well, it is true that they can vibrate and we can think, for example, of the strings of a musical instrument such as a guitar. They are closed forces, like a closed ring. It is possible that these closed rings, the strings, which perhaps are very heavy, are out there in space. We haven't discovered any yet, but they are solutions to mathematical equations and they could be there. Unlike the strings of a musical instrument, those out there are not made of atoms.

E.P. They must be made of something.

E.C. They are continuous matter that forms a string without empty space between the particles. Just continuous matter.

E.P. So, we need to rethink the different states of matter that we know. Let's see: the Greeks believed that the world was made of air, earth, water, and fire, right? And physicists like you now say that there are gases, liquids, solids, and plasma. We'll talk about plasma later. It seemed normal to us, so familiar with water, that a drop in temperature leads to a change in the state of matter from liquid to solid, for example. But the world has become more complicated since the defenders of complexity and chaos theories like Steven Strogatz [see chapter 14] or Albert-Laszlo Barabasi discovered that what appeared to be normal, very banal—I mean the change from liquid water to the solid state of ice—is, in reality, a phase transition that takes place in self-organizing systems. And we still don't know how to really explain the mechanisms by means of which the water molecules suddenly give up their individual lives to form a compact, immobile collective like ice. Of course, it is fascinating to think of water vapor molecules that move independently banging into each other; and of the same molecules which, in their liquid state, come together randomly and promiscuously—as if they were simply holding hands. And, finally, the same molecules decide not only to hold hands, but to cling together, inseparably, as if in a sacred rite of mutual devotion. And, as if this weren't enough, now you tell me that over 99 percent of the universe is "plasma." The people in the street don't really know about plasma. We understand what gases, liquids, and solids are—but plasma, what is plasma?

E.C. We in physics know well what plasma is: it is a state of matter in which, like electrons and protons, particles are not connected to each other as in atoms, but they exist as individual objects. The state of matter called plasma exists at very high temperatures, thousands or perhaps millions of degrees. At these temperatures the atoms break and individual elementary particles are free in space. That is plasma.

E.P. "Plasma" is the state of matter inside the sun?

E.C. Correct, it's the sun.

E.P. What are we made of?

E.C. It's almost certain that no plasma is inside us; our temperatures are far too low. But we do have matter in the other states: fluids, like blood; solids, like teeth; and gases, such as the air we breathe.

E.P. Are there other states of matter that we don't know?

E.C. There might be. For example, for over ten years now physicists have been studying intensely a certain mystery of astrophysics, so-called dark matter. The only way to explain gravitational forces and the evolution of the universe—the attraction between different objects—that experiments demonstrate is that the mass of the universe is ten times greater than what we observe. Therefore, on the earlier assumption, there are substances we cannot see that make up most of the universe, the famous dark matter. And we don't know what it is because it interacts very weakly with other kinds of matter that we do know.

E.P. Is it true that we walk through a cloud of electrons? I mean, when two solids come into contact with each other—when we shake hands, for example—what is it that is shaking your hand with mine, a cloud of electrons? Do your atoms and mine remain separate?

E.C. Yes. The simplest way to imagine an atom is as electrons that revolve around a nucleus, like the satellites that revolve around the Earth. Electrons have a negative electric charge; that is, they repel other electrons. Therefore, when we shake hands or touch something, the electrons in my hand are repelled by the electrons of the other body. So, because it is rejected by the other electrons, my hand can't really go through your body.

E.P. Otherwise, the ground would swallow us up.

E.C. Of course.

E.P. It seems incredible that we walk on clouds of electrons. We are not stuck to the ground because, as you say, if we didn't have electrons between us and it we would go through it.

E.C. That is absolutely correct. The electrons in the ground repel the electrons in our shoes and that is why we walk, as you said, on a cloud of electrons.

E.P. Before, you mentioned strings, perhaps very heavy strings somewhere in space. But is there any similarity between the world of solids on Earth and that of strings?

E.C. That's a good question. Almost all the physics of the universe, of strings or whatever's up there—such as the monopoles, which are very strange objects—is described by the same mathematical laws as solids. In superconductors, for example, are vortices, and those vortices resemble the cosmic strings we talked about earlier. We use the same mathematical models, so when we study superconductors, both theoretically and experimentally, we learn about the physics of cosmic strings. This is just one of many examples. There are hundreds of other examples in which phenomena that we observe in solids are incredibly similar to the phenomena that we perceive in certain cosmological processes. Thus, by the study of solids, we can discover the universe in a cheaper way than by building enormous particle accelerators or by sending ships into space.

E.P. When we talk about solids, what are we talking about really? Metals, wood, or new materials? Can we create new materials?

E.C. We refer to all kinds of solids. And we create new materials all the time by combining different elements at different temperatures and pressures. Before creating them, we carry out simulations to gain an idea of what we will obtain by the combination of elements. Starting from the idea and by means of experimentation, we create many new materials.

E.P. How lucky we are to have the intellectual ability to simulate, the way a pilot simulates what will happen in a stormy sky! But when we say, "We are capable of the creation of new materials," what are we saying? Can we create molecules that did not previously exist? Life?

E.C. That is a very good question. Without a doubt, biophysicists and biochemists create new molecules, even very complicated ones, but the creation of new life, no one can do that. Let's go back to your first question about whether what we see outside ourselves really exists. Yes, we really see what is out there, but it is surprising because, for example, the trees we see outside my window, those trees die and grow again. The most interesting question in science, in my opinion, is to know why all that happens, and why in the physics, biology, or chemistry laboratories we can't reproduce the life process. We don't know how to repeat the process of biological evolution in which the atoms and the molecules come together and, above a certain size, start to reproduce. From the point of view of a physicist, what we see outside the trees outside my window that die and grow again, remains absolutely astonishing.

E.P. This reminds me of a conversation with Stanley Miller some years ago. He appeared on the covers of the most prestigious magazines in the 1950s because he obtained synthetic amino acids in the laboratory, the building blocks of life. I asked him: "Why did you not continue creating proteins and then life?" "I thought I could do that the week after, but have been trying for forty years and I still haven't managed it," was his reply. Perhaps Stanley

Miller should have thought of what you've just said, that there's something very surprising in this process that organizes itself, as if there were a predetermined objective. Why does that tree grow and grow and yet we cannot repeat it in the laboratory. All we can do is germinate the secret written in its seed—plant it.

E.C. In the course of the past three billion years, evidence has emerged on Earth that evolution goes in a single direction. Life acquires greater complexity. Order increases on the surface of the Earth. Materials are very ordered inside the trees, the humans and other animals, the buildings, tables, and books. Whereas three billion years ago there were only gases and solids; everything was disordered and chaotic. And this process is moving in one direction. Although we can't explain it, nor simulate it in the laboratory, greater complexity as we come toward the present seems to be what happens.

E.P. Eugene, for now we'll just accept that the external world probably does exist.

E.C. That's a good supposition, especially for me as a physicist, because otherwise I would be out of a job.

E.P. comments: In that conversation with Chudnovsky in his Lehman College office in the Bronx, we didn't have time to try to decipher one of the greatest paradoxes of science when it comes to ascertaining what is going on "out there." The reader will have noted that the undeniable belief in increasing order among the things around us clashes with other undeniable evidence, which is that posited by the so-called second law of thermodynamics: the spontaneous evolution of an isolated system tends toward disorder and is always translated into an increase in its entropy.

It's not true—Stephen Jay Gould, the great paleontologist, repeated to me in his office at Harvard University, half junk room, half laboratory, with an old dilapidated armchair in the middle—that we're moving toward something increasingly bigger and more perfect.

And when Ken Nealson himself, one of the most profound and precise astrobiologists on the planet, declares that "life is a mistake" (see chapter 24), I think that both of them are under the legitimate influence of the paradox I referred to before. The spontaneous evolution of an isolated system is always translated into an increase in its entropy.

The word *entropy* was used by Clausius in 1850 to describe the degree of disorder of a system. Therefore, the second law of thermodynamics states that isolated systems tend toward disorder.*

Living organisms, among them humans, like all else in the universe, tend to decline, deteriorate, become more disordered, and disappear since they are systems of unstable dynamic balance, that is, their entropy increases.

Life, its maintenance, is equivalent to a reduction of entropy, as seen in the phenomenon of homeostasis that depends on a continuous flow of energy. Life is a phenomenon composed of open or continuous systems, capable of reduction of their internal entropy by expenditure of either nutrient substances and substrates or free energy that they take from their surroundings, returning it in a degraded form.

The second law of thermodynamics condemns all systems to degenerate, decay, rot, and die. The pavements crack, houses age, consecutive translations alter original meanings, stars wink as they go out.

Brian Swimme and Thomas Berry suggested the cosmogenetic principle, which allows us to see the evolution of the universe in a more real way, by considering this principle as a complement of the second law of thermodynamics. It refers to the construction of order, the increase in complexity and therefore in information. According to this principle, the second law of thermodynamics is a statistical phenomenon, so that in certain conditions, the increase in disorder can be reversed.

In the theory of communication or of information, entropy is a number that measures the uncertainty of a message. Entropy is zero when the certainty is absolute. When we add information to a physical object, what we are doing is ordering the elements that comprise the system of that object in a particular way.

Biological and economic systems are not isolated systems, as both receive the heat of the sun. Therefore, as long as they receive more energy than they emit, economic and biological systems can reduce their entropy. In plain speech, as long as there is usable light energy from the sun, biomass can increase and the world's gross national product can increase, as well.

It could be said that a healthy living being, a company, or a chugging locomotive have low entropy; they contain information. If the disorder in the components of the individual, business, or machine increases, its entropy increases. There is a certain threshold, a certain level of entropy above which

*But living systems are neither isolated nor closed. They are open to matter, and energy always must flow through them. See chapter 27 and Schneider and Sagan, 2004 on the purpose of evolution.—LM

the living being dies, the company becomes bankrupt, and the machine stops working.

Living beings, companies, or machines are not isolated systems. They can use energy provided by other systems to correct the disorder, that is, to reduce the entropy. However, the cost of such an effort is so huge that it is legitimate to say, as Ken Nealson said, that "life is a mistake." From experience, we know that this possible intervention has a limit. Until now, we know of no living being, company, or machine that has lived or functioned eternally. Except life as a whole system from the time it began.

— 32 —

Atomic Consciousness

Interview with Heinrich Rohrer

> A molecule recognizes another one or it fails to recognize a different one, but it is not thanks to fantastic equipment.
> —Heinrich Rohrer

We belong to the world of small things. Perhaps because of that, it was not coincidence that first the microscope was invented and only later the telescope, which Galileo was immediately able to use to look at the large celestial bodies. The great milestones in penetration of the world of small things were because of the microscope, including the discovery of bacteria. Afterward specialized medicine and biology were truly created. These were followed by the electronic and atomic microscopes. The last barrier that prevented us from

---

Heinrich Rohrer won a Nobel Prize in Physics (1986) for his contribution to the design of the scanning tunneling microscope that allows visualization of subatomic scale detail.

seeing the fundamental particles from which the universe is made was broken down thanks to the scanning tunnelling microscope (STM), associated with Heinrich Rohrer, who in 1986 received the Nobel Prize in Physics for his work in the IBM laboratories in Zurich. The tunnelling microscope not only makes it possible to observe the world of molecules and atoms but also to process the information at that level.

Eduardo Punset: Let's consider models. The American physicist Richard Feynman said: "I want to reach atoms and to see them." If we discover how one atom relates with another—and we are made of atoms, and molecules—then we will know what nature is. Is that true, or do we still not know what nature is?

Heinrich Rohrer: It is very complicated; nature is complicated. Knowing the atoms does not enable us to understand the operations of complex systems composed of atoms—though I think that atoms are basic. We need to understand how they behave in certain circumstances. The new challenge of nanotechnology is to learn more about systems composed of atoms that can be used for very specific purposes. But it is a very great leap from atoms to all of nature.

E.P. Don't we need some kind of self-assembly and self-organization of atoms? How do we get them to self-assemble and self-organize?

H.R. That is very ambitious. We do not analyze things that self-assemble because they would stop self-assembling. We study certain structures and why they behave as they do. If we want to create a system we cannot imitate nature. Nature has had a million years to think about self-assembly. If we want to assemble something, we cannot wait for nature. We must collaborate on the assembly. This is one of the great current challenges, because we understand a lot about self-assembly and its mechanisms, but if we want to form a system, we have to be capable of generation of a self-assembly process. But we still don't know how to do it. Perhaps it is possible in very crude circumstances. Certain experts say that carbon nanotubes, instead of copper wire, can be used as the base connectors between one transistor and another. This may be true up to a certain point, but it solves nothing if we can't create the carbon tube in situ, if we can't put millions of carbon tubes in the right place. We have to start a process of growth or assembly in the specific place so that it develops the way we want. One of the great challenges of nanotechnology, material science, and nano-material science is we still are very far from that. Science needs great challenges.

E.P. It is irritating for a scientist to see that molecules assemble by one attaching itself to another, and another one to yet another, until they form a complex system; and given that we now know the basic nature of molecules, made of atoms, and thanks to the STM microscope and others, we can identify images, yet we are not capable of instigating these self-assembly mechanisms? We will be in the future, no?

H.R. We probably will never depend exclusively on self-assembly. Perhaps we will create models by miniaturization, which is how we work now in microelectronics. We design many nanoelectronic processes. Probably we will use some hybrid processes in which we will make a miniaturized framework by lithography. On these models we start to develop, by means of self-assembly, the smaller details.

E.P. Because when we think of the self-assembly and self-organization of atoms, is it with the intention of creating certain products? Or for the purposes of scientific research? What kind of products do you mean?

H.R. In nanoelectronics many are getting close to electronic circuits, new components beyond transistors or different types of transistors. Nanotechnology will probably be a field in which connections are made between the virtual world of data processing and the real world of action. Nanotechnology will provide some fantastic sensors. All processing requires input, and something must be done with that input to create an action. At present there is a lot of processing, communication of the same thing. In the future there will be a point of contact with the real world in both perception and action. That will be the place of technology. For ten or fifteen years we will still be able to do what we are doing now but at a smaller scale. Now, when we talk about a small circuit, we still think of computers. That is not necessarily the area in which nanotechnology will be optimized. The world of the processor must be connected to the real world. One of the most interesting characteristics of nanotechnology is the size of molecules. The molecules in our body work such that one molecule recognizes another one. Not thanks to fantastic equipment: one molecule is used to recognize another. And one basic principle of nanotechnology is to recognize small things with small things. With the scanning tunneling microscope (STM), an atom, if it comes sufficiently close, recognizes another atom in the needle. That is the principle of the STM: one atom is recognized by another. This works in several ways. By the induced current, as in the tunneling microscope, or by joining in different ways on the different types of molecules. We create an individual molecular sensitivity: a key development of nanotechnology.

E.P. Where does this lead; can it be applied to biomedicine?

H.R. It can be applied to any field. I imagine myself carrying a very small sensor that immediately perceives that someone near me has the flu. In the end it would stop being a problem. I think that we will be capable of such recognition, but we also have to think about the consequences. We talk about private information we can gather. People no longer have private lives; we know everything about them. A possibility opened up with computers. But imagine if we were to immediately know a person's state of health when we meet him. We will have to decide: How should I react if I know that a person has AIDS? Is there time to isolate him or her? This is a much more important, more difficult question than, for example, data privacy.

E.P. You write that nanotechnology reveals both great prospects and great fears.

H.R. Despite the fears, we must really think about this subject because we need to handle consequences properly. If we observe the progress of science and technology during the last two hundred years, the world has not been destroyed yet. People who proclaim the destruction of the world, who fear the end of mankind, exaggerate. We need to be confident that people become not only more intelligent but more sensible. Sometimes we are assailed by doubts.

E.P. An interesting aspect of your work is your insistence on a multidisciplinary approach. The question is no longer microelectronics or nanoelectronics, but rather general molecular processing. Whether in mechanics, quantum physics, electronics, or biology, Heinrich, if we analyze animals' intelligence, we discover that it is intelligence by association. You just alluded to the capacity of association through information, simultaneously to associate one thing with another. Colleagues like Nicolás García say that intelligence will arise from the enormous quantity of information. Do you agree?

H.R. Yes. We need a lot of information for intelligence. A little child does not yet have intelligence, but he or she gathers information and stores it. He or she develops in a certain way, and so does technology. One challenge is how we can move forward. I don't know whether machines can do creative work, but we will build more active machines than at present.

E.P. When you go down inside to that scale, to that limit, enter the quantum world, matter behaves very differently. The same particle goes toward two different poles! Does this pose a real problem?

H.R. Probably one of the great challenges of nanotechnology is to exploit quantum mechanical behavior better at the microscale or the macroscale. Now we already exploit quantum mechanics in a more obvious way than the macroscale. Each atom behaves in a quantum manner, that is, quantum mechanics is inherent in matter. But the quantum behavior of the properties of any matter is not obvious. A large stone is thrown and we calculate its trajectory with normal Newtonian mechanics; we don't need quantum mechanics. Yet at the very small scale, quantum mechanics plays a critically important role. If we have a very thin wire and a constriction point, which is what Nicolás García investigates, then transportation is of a quantum nature, for example.

E.P. Does not the control of condensed matter, both living and dead, one atom after another, open up a wide range of possibilities? When we think of of ribosomes of living cells, which build proteins according to instructions of RNA, once we develop the ability to control atoms, one after another, can we imagine analyzing the atoms of the RNA instructions to change them? Am I just silly? Can we change not just nature, but also the instructions that nature uses to build organisms?

H.R. My colleague is convinced that the laws of physics themselves may be a sign of spirit or intelligence. Everything is composed of atoms, but is the human way of thinking simply a chemical process in the brain? Is it something more? If it is just chemistry then we have to allow the process to be intelligent. Now we identify intelligence, living matter, and inanimate or dead matter as separate. That separation probably is not very reasonable. Nanotechnology brings together biology, physics, and engineering. Perhaps someday our perspective on these three categories— intelligence, living matter, and dead matter—will differ. If everything is more blurred in the future, we might have a more reasonable transition from one to another. Who knows?

Too Huge for the Atom, Too Tiny for the Star

Interview with Sheldon Lee Glashow

> Imagine it: a full moon all the time in Miami. But suddenly it comes closer and explodes into millions of burning fragments! It is easy to predict. The past is less interesting to me, because it has already happened.
>
> —SHELDON GLASHOW

Of course, by definition, we are in the middle.

It's logical for us to ask the question of whether the universe exists, as Nobel Prize winner in physics Sheldon Lee Glashow said to me in the course of a visit to Barcelona.

Sheldon Lee Glashow, Nobel Laureate in Physics (1979), is the Metcalf Professor of Mathematics and Physics at Boston University.

We are too large to see tiny things and too small to see the galaxies. We have always liked being in the middle. In ancient times, the Earth was the center of the universe and Jerusalem was the center of the Earth. We also like to think that in the world of dimensions we are in the middle. We are. There are things that are millions of times smaller than us: atoms, nuclei, quarks, superstrings, going down the scale. But also with larger things: Earth, the solar system, the Milky Way and, of course, thousands of millions of galaxies and the universe as a whole. We are in the middle. We stand between cosmology, the study of the large, and particle physics, the study of the smallest things in the world. And between these two extremes we have the common sciences: chemistry, biology, geology.

Sheldon Lee Glashow resumes the conversation where Eugene Chudnovsky stopped when he boasted that he did not need spaceships in order to study the universe, he had only to look at greater depth into solid-state physics.

Sheldon Lee Glashow: We speak about two extremes simultaneously: on the one hand, the universe, and on the other the smallest particles. But we have discovered that the study of particles is very related to the universe con-

Table 2. Units of Measure of Lengths and Distances.

UNITS	METERS	EXAMPLES
	10^{-18}	diameter of electron/quark
Femptometer	10^{-15}	1 fm, radius of the proton
Picometer	10^{-12}	50–600 pm, atomic diameters
Nanometer	10^{-9}	0.1 nm, atomic hydrogen diameters
Micrometer	10^{-6}	1 μm, diameter of a bacterial cell
Millimeter	10^{-3}	9 mm, 5-week-old embryo
Meter	1.0	1.7 m, average height of people
Kilometer	10^{3}	8,846 km, height of Mt. Everest
a thousand kms	10^{6}	6,371 km, radius of the Earth
a million kms	10^{9}	0.7x106 radius of the Sun
Light-second (1-s)	300,000,000	499 light-seconds, orbit of Earth
Light-hour (1-h)	108 x 10^{12}	10 light-hours, radius of solar system
Light-year (1-y)	946 x 10^{15}	4.3 light-years, distance to the nearest star
		105 light-years, diameter of the Milky Way
Parsec (pc)	3085 x 10^{16}	radius of the visible universe

Measurements that help us to see our relationship to our surroundings.

ceived of as a whole. In practice, the large follow the same series of rules as the small. Fundamental physics is applicable everywhere, independently of the size.

Eduardo Punset: Then you don't think there are two physics: cosmology, at the large end and particle physics on the small.

S.L.G. No, there are two names: cosmology, which studies the whole universe, and particle physics, which studies the very small. But they are related, they always have been. Consider the time of Galileo in the quite distant past (1564–1642). He was a very intelligent man, but he was not always correct. He discovered that when very little water remained in a cistern, despite the use of a water pump, he could not raise the water to a height of more than nine meters. Galileo did not understand this. He attributed the mystery to the claim that a column of water nine meters high would break, as happens with a chain too long to bear its own weight. He never understood. A student or colleague of his, Torricelli, resolved this mystery when he realized that it was not the pump that made it possible to raise the water. Rather it's the atmospheric pressure. The weight of the surrounding air determines the maximum height of a liquid in a recipient independently of the power of the vacuum pump. Torricelli wrote: "We live under an ocean of air." This cosmological affirmation comes from the seventeenth century; the air is down here and not up there. The concentration decreases as we go up, until it disappears. The world is surrounded by a vacuum. The Earth is surrounded by air, by the ocean of fine air that we call the atmosphere. This cosmological discovery enables us to explain small things: how we breathe, how a baby sucks milk. Phenomena such as the nature of pressure are cosmological discoveries related to large things. And later came Sir Isaac Newton (1643–1727). Everyone has heard of Newton. An apple fell on his head and so he discovered the laws of mechanics, that is, the laws that apply both on the earth and in the sky. But did the apple explain how the Moon moved, or did the observation of the Moon's movement explain how the apple fell? Do you see? The world of small things, like the apple, and the world of the large, like the moon, do follow the same rules.

E.P. When we think of biologists who see bacteria through microscopes, we think they feel certainty because they are supported by the knowledge of chemistry. And chemists, in turn, in their world of molecules, trust the validity of their work because they have the backing of physics: of phenomena related to protons, neutrons, electrons, and quarks. But what backing has the physicist? Where do the quarks come from?

S.L.G. Yes. Biologists work with what they see. Sometimes, they are large objects like elephants, sometimes they are small ones such as DNA molecules; or they are the atoms that make up these molecules. But atoms mark the limits of size: as we approach the very small, chemists can no longer give us adequate explanations. Chemists study the atoms as structures that constitute different realities, but physicists analyze what is inside the structures and how they work. There are several levels in physics. For instance, atomic physics. What do atomic physicists do? They study the composition of the atom, its nucleus, and the enormous number of electrons that revolve around it. People ask, what are they looking for? Do they want to know why the stars shine? What is radioactivity? The answers are given by another level of physics known as nuclear physics. Some physicists explain how radioactive substances are made for medical purposes, or how nuclear scanners that enable us to image the inside of the human body to show how it works. Beyond the naked eye they reveal the nature of disease. However, such nuclear physicists don't ask themselves "What is a proton or an atom?" They simply don't question; like chemists and biologists they simply assume the existence of atoms. Then come particle physicists. They do want to know what is inside an atom, its proton and neutrons. Thirty years ago, they discovered that protons and neutrons are composed of two types of quarks, but there are other particles that are strange.

E.P. Do they behave differently?

S.L.G. They are strange!

E.P. Is that because some of those particles can be in different places at the same time?

S.L.G. All particles can be in different places at the same time. That is not what is strange about the particles. In reality they are not so strange. They are composed of a third type of quark. Why does the strange quark exist? We don't know yet, but now we know that there are not only three. A fourth type of quark was discovered in 1974, a fifth in 1976, a sixth type of quark in 1994. They are different classes of fundamental particles. The mystery is that differences between particles exist. When I mentioned the word "strange" from your question I suspect you thought of quantum physics. Yes, quantum mechanics is a very strange discipline, but its strangeness differs from that of quarks.

E.P. If biology is backed by chemistry and chemistry is backed by physics, what is physics backed by?

S.L.G. The common sciences like biology are very specialized. What does biology study? Living beings, according to recent investigations, appear only to be found on Earth. But our planet is only one of thousands of millions of planets that fill the universe. That is why I think biology is a specialized discipline. The degree of specialization is even greater in psychology, because it studies the human mind. Or your field of study, economics, which is an even more specialized discipline whose subject is money! Physics, on the other hand, encompasses everything—well, perhaps not money, but certainly everything else material! Physics encompasses chemistry, biology, astronomy, cosmology. Everything is based on physics. All sciences are, in the end, physics.

E.P. Let's talk about something very close to both of us. I refer to the four divine creatures that govern everything, the so-called four forces: gravity, the electromagnetic force, and the two kinds of atomic nuclear interactions, strong and weak. Apparently, the four forces were originally unified when the universe began twelve billion years ago.

S.L.G. Perhaps . . .

E.P. Perhaps, but what is certain is that they are now different from each other. One is transported by photons, another by protons. From when does the separation of these forces date?

S.L.G. We only know part of the answer. Let me return to this argument. When I speak of four forces, or you say, "there are four forces," it reminds me of my fifth grade teacher who told us there were seven continents.

E.P. Wasn't he correct?

S.L.G. I had never heard of Australia. I certainly never knew the difference between Austria and Australia when my teacher told us there are seven continents. This is the wrong way round. I think we should have begun by learning all about the continents. Probably the best way to teach us would have been by showing us photos of the animals that live in Australia, of the customs of the original Australians. You, and people in general, can't know what physicists mean when we say "four forces." People only know two; biologists, chemists, or bank employees know only two forces: if you drop an object it falls to the ground because the "force of gravity" exists. And then there is "electricity."

E.P. That's true.

S.L.G. Most everyone knows that the force of gravity exits. And when they go to school they learn that what makes matter, leaves, tables, is electromagnetism. What we see, hear, or touch is electromagnetism. Right now this shaft of light that illuminates is also electromagnetic. There's nothing else.

You are talking with me at the same time as you examine electromagnetic radiation emitted by the surface of my body. Therefore, most of science, and of human life in general, relates to gravity and electromagnetism, which make atoms and things work the way they do. But what are the other two interactions, the two other forces to which you refer? They only exist inside the nucleus. Normally they can't be seen. The nucleus is composed of protons and neutrons, as I mentioned, held together by a kind of glue that constitutes, or forms a part of the strong nuclear interaction. That nuclear force in reality holds quarks together to create protons. The fact that they are united is only related to the atomic nucleus so it's pretty irrelevant.

E.P. To daily life?

S.L.G. To daily life, to nothing! Beyond the energy of nuclear bombs, nuclear medicine, the functioning of the Sun, or the activity of the stars, which are related to atomic nuclei.

E.P. If I understand you correctly, the weak nuclear force contributes to maintainance of the thermal combustion of the Sun.

S.L.G. Yes.

E.P. And given that it is weak, it enables the sun to last for thousands of years. Otherwise the Sun would blow up.

S.L.G. This is interesting to me. The Sun is fascinating in that it uses all four forces, not just one. Gravity holds it together. The pressure inside the sun is very high and therefore so is the temperature. The electromagnetic force is required. Without it the sun would be like a bomb that explodes. Strong interaction is also required, because without it protons would not join to form more complex nuclei. No source of nuclear energy would exist without the strong interaction. As you said: we need the weak interaction that makes it possible to transform protons into neutrons. If this force didn't exist, quite possibly there would be no sun.

E.P. There would be no energy, or therefore life, without the strong interaction that holds together the protons, no? Is there any danger that this all may fall apart someday?

S.L.G. Do you mean, "Will everything continue to hold together?" We don't yet know, although yes, we are almost certain that it will. The experiments that were carried out demonstrate that protons live longer than we thought. Experiments in Japan and the United States suggest we shouldn't worry because the protons and neutrons will last, at least, for 10^{30} more years (this universe is at most 10^{14} years old). That's a long time, a very, very long time.

E.P. Apparently, then there is no risk that everything will suddenly fall apart, no?

S.L.G. No. However, it is fantastic, one of the marvels of modern physics that we can witness both the appearance and destruction of particles at the same time. Two aspects that should be noted: two answers to questions that lead us to reflect. Why are atoms that are here on Earth the same as the ones there in space? Salt that is obtained from the sea is the same as the salt that comes from mines. You will tell me that in both cases it is sodium chloride composed of sodium and chlorine. But why are atoms of sodium the same independently of their origin? You will tell me that sodium is composed of the same number of protons, neutrons, and electrons, so different sodium atoms are in reality identical. But why are all protons, all electrons identical?

E.P. Perhaps because everything began at the same instant, they started from the same . . .

S.L.G. Well, not exactly. What we call particles don't exist. A particle is a manifestation of a field, and there is a single field that describes all the electrons. It's an object, a mathematical object that creates and destroys particles, all the same! The forces of nature are due to the exchange of bosons, the so-called force particles. The strong force, the most intense, does not affect the macroscopic world due to its short reach. Gravity, on the other hand, although it is the least intense force, is the best known. This force keeps our feet on the ground and the planets in movement around the sun. The much stronger, the electromagnetic force, binds electrons and nuclei inside atoms but also binds atoms to form the molecules of solids or liquids. The weak force is responsible for the disintegration of neutrons and other particles, but also for some interactions in the centers of stars. The strong force confines quarks to the protons and neutrons, and acts against the electrical force that repels them. Thus the strong force keeps protons together in the nucleus.

E.P. Why are the particles all the same? The second issue that amazes you is creation and destruction.

S.L.G. It is the end of the vision of Democritus who, as you know, first suggested that the universe is composed of eternal atoms. We now know that atoms are not eternal. Rather they can be created, made, and destroyed by processes or energy. When Einstein's theories of relativity and Heisenberg and Schroedinger's quantum mechanics ideas were combined, the possibility of creation and destruction of elemental particles arose. And it is happening!

E.P. Does this revolution in our knowledge of atoms also affect our knowledge of the four forces? Do we know how and why they separated?

S.L.G. We know, or think we know, that there was a time when the interactions were united. Certainly the electromagnetic force and the weak nuclear interaction were united. But what mechanism separated them? What made them differ? We will have to answer this at the start of the twenty-first century, thanks to the experiments that are being carried out at the European laboratory for particle physics in Geneva. The new particle accelerator has been designed specifically to find what we call "the fifth interaction," to learn what separates electricity and magnetism from the weak interaction and generates mass. We return to the start of our reflection: why do particles have mass? In the beginning apparently nothing had mass, everything was like photons, particles without mass. They later acquired mass. The protons and the electrons have mass they acquired at the dawn of the universe. We want to know why, how, and where did these elements acquire mass? I don't know.

E.P. Perhaps it's easier to take a look at the future.

S.L.G. Yes, for us it is an easier question to answer than one about the past. We now know that certain phenomena will take place. We now know that although we may live in peace on Earth for several thousand million years more there will be disasters when the Sun explodes or the Moon collides with the Earth. The Moon is moving away from the Sun, but the day will come when the Moon remains immobile in the sky over Miami. Imagine it: a full Moon all the time in Miami. But suddenly, it comes closer and explodes into millions of burning fragments!

E.P. Fascinating . . .

S.L.G. Yes, it is fascinating, but it's not good news for us. Though by then we may be orbiting around another star where we feel safe. Perhaps that future will take thousands of millions of years to arrive, but it is easy to predict. The past is less interesting to me, because it has already happened. To know how it happened is a historical problem. We learn about the origin of life and of animals in biology. We know that life began at a very remote time in the history of the planet.

E.P. When matter appeared.

S.L.G. That's right. It took a few billion years, but then it rapidly flourished. The universe was already about ten billion years old, very old, when the Earth formed. We know some things about the first phases of the universe. We know when atoms formed. We use a linguistically abominable term, *recombination*, to describe the first formation of atoms. Recombination? No, it was the first time they combined!

E.P. Nucleus, quarks, and electrons?

S.L.G. Nucleus, quarks, and electrons. That is an atom. They combined when the universe was only three hundred thousand years old.

E.P. And light first appeared at that time?

S.L.G. Yes. Before that precise instant the universe was a dark place. Suddenly this marvelous universe appeared!

E.P. I would like to have seen that instant. . . .

S.L.G. I don't think you would. The temperature was probably the same as that at the surface of the sun. It was not the best time to be there. There was really no "place" at that time. No structure or place to be. No planets existed, nor stars, nor structures: only light. Virtually the whole universe was identical. Experiments recently performed explain how the universe changed from a burning, uniform, homogeneous place to a structured, very cold volume full of interesting things like planet Earth. That is what happened!

E.P. comments: If the truth is that at the start of any creative project there is an emotion—a commonplace idea that it took the scientific community many years to accept—the overriding emotion of physicists is related to vertigo: they are poised on the point of discovery of first phases of matter. Their vertigo affects everyone. Walking in winter around the CERN campus, in Geneva, full of frost and sleet, I identified that emotion in a young Spanish scientist.

Miguel Angel Sanches, a researcher from the Valencia Particle Physics Institute, talked about one of the great hypotheses of recent theoretical physics—the existence of extra dimensions.

We humans live in a four-dimensional world, the three dimensions of space plus the time dimension added by Einstein with his theory of relativity. Our normal world can be described as movements in the four dimensions. But apparently when we try to explain the four forces that govern the universe we realize that four normal dimensions—x, y, and z of space, plus time—do not suffice.

Miguel Angel, like many other physicists, accepts the existence of extra dimensions, those we can't see but we need mathematically. They permit, in principle, the unification of the forces of nature. The additional dimensions help us understand the four forces as manifestations at different scales of a single force.

Gravity, the weakest interaction and paradoxically the first to be identified, is critical in the search for extra dimensions. This force, gravity, could spread through the other dimensions, which would explain why it is so weak. Gravity is dispersed in gravity dimensions we don't see; it is not confined like the other three forces to our three-dimensional world.

If gravity is formulated in quantum language, "the graviton" is the particle that carries the force of gravity, just as the photon of light—with no mass—carries the electromagnetic force. This "graviton" has not been observed, Miguel Angel continues, perhaps precisely because it is spread through the extra dimensions. In the huge hadron collider, in experiments planned to begin at CERN in 2007, they expect collisions between protons at very high energies to occur. If, in these collisions, any graviton is produced, it will be detected, although not directly. Presumably it will disappear into extra dimensions! The trace any gravitaton leaves on the detector (seen as a spray of particles that lack balance of energy and momentum) will be measurable!

Such a finding is thought to lead to "new physics," which, unlike previous discoveries, would not contradict the "current standard model" of particle physics. Rather a "graviton" discovery, it is anticipated, will extend the standard model, although in a simplified way, to fundamental principles. Apparently the current standard model, despite its capacity to describe most processes and predict new ones, leaves too many questions unanswered. The discovery of new dimensions and the quantification of gravity would be the first step toward the unification of all the forces. Such a successful set of predictions would correspond to one of the unified ideals of all physicists, and relate to philosophical descriptions expressed by William of Ockham in the fourteenth century. This Scottish philosopher wrote, "entities should not be multiplied beyond necessity, the simplest hypotheses are always to be preferred." We agree.

Not only physicists, but biologists, neuroscientists, robotics specialists, and those who work on artificial intelligence share this feeling that any day—or in a restful and unconscious learning dream during the night—science will unexpectedly provide an answer that explains everything. The sensation floats in the air in the international scientific community. Soon we will realize a then totally obvious but ignored truth, say, about the long-sought secret of the origin of matter and therefore of ourselves. Why have we not already realized it? We may need a new Democritus, Galileo, Darwin, or Einstein.

The new discovery, I anticipate, will result from the unfettered knowledge that, like the four forces that move the universe, will one day unite the disciplines. They will return—just as before each one went off on its own way, all was together at the origin of the universe. This reunification, this recombination, already happens at the best research centers on the planet.

Another requirement for that leap forward in our understanding of matter will consist of the consolidation and dissemination of the virtue of humility, inherent in science. Unlike dogma or "revealed knowledge," science is subject to verification, experimentation, testing, and severe open criticism. The advantages of scientific methods, observations, measurements, consistency of evidence, etc, compared to knowledge by revealed authorities—mostly unfounded fairy stories—is that the latter unwittingly led to "scientific fundamentalism." This dogmatic stance is comparable in its devastating effects to religious fundamentalism. Certain writings by Carl Sagan come to mind (Sagan, 1996). Particularly, I recall the vehemence of a Nobel lauriate winner—he who received one of the most well-deserved Nobel Prizes since Alfred Nobel created them in his will in 1895—against homeopathy. Many examples, attitudes toward astrology, continental drift, cause of gastic ulcers (see chapter 29), of scientific dogmatism, still exist.

I never forgot my meeting with the physician and researcher Jacques Benveniste near his former laboratory, some fifty kilometers from Paris. He had just taken refuge with two or three loyal assistants in a shed with an asbestos roof, near the modern laboratory he directed until the moment when, confronting the Medical Association, he maintained against hell and high water that water itself could remember, had memory. According to homeopathic experts, even if drugs are diluted to the point where they no longer contain the active principle molecule, they continue to be effective. The memory properties were ascribed to water itself!

With current knowledge, it is still very risky to claim that the millions of patients who resort to homeopathic remedies recover. If they do, is it certain that recovery is due to other factors than the "placebo" effect? Experiments suggest that homeopathic remedies are false, and that other nonplacebo-effect factors may eventually be identified, that there is a physiological explanation beyond the impact of the mind on metabolism. Science keeps channels of communication open.

Under the Ancien Régime, in France, prostitutes were marked with a fleur-de-lys on their shoulders. For Jacques Benveniste, that the reductionist scientists, in the same way, comparably labeled homeopathy is disgraceful. They refuse to admit that "a substance diluted in water has been so diluted that in reality it has disappeared, nothing remains. . . . But that, in reality, there is something, a kind of footprint, a memory, that would explain why homeopathy is effective." They maintain that for curative activity molecules are needed, however virtual the world of information may be. Benveniste replies: "When you play a CD of a singer or an orchestra, the ear doesn't distinguish the source. If you play a CD of Luciano Pavarotti's music, the ear doesn't know if it hears a recording or Pavarotti himself. The Nobel Prize winners who criticize me believe that molecules must be present for curative activity. Theirs is no more than an article of faith."

Much more tragic precedents exist in the history of contagious diseases than the official establishment's refusal to accept homeopathy! The physician Philip Ignaz Semmelweis (1818–65), in the 1840s, wrote that puerperal fever, the cause of death of many women in childbirth, was mainly due to poor hygiene and infection. Before then, puerperal fever was considered to be a disease inherent in childbirth for which nothing could be done. But Semmelweis's investigations demonstrated that in delivery rooms attended to by midwives, maternal mortality was no more than 1 percent. However, in those attended by students who had come from the autopsy amphitheatre, deaths were in excess of 30 percent. Semmelweis concluded that the rotting matter present in autopsies was the cause of puerperal fever and obliged everyone who entered a delivery room to wash his or her hands with chlorinated water.

Though these measures prevented puerperal fever, the majority of doctors rejected Semmlweis's thesis. Thousands of women continued to die needlessly in childbirth for decades, in untold pain. Semmelweis was expelled from the hospitals. He was accused of being an enemy of the Austro-Hungarian Empire. He ended his days in a lunatic asylum, suffering from feverish delirium, perhaps from septic anemia after he cut his finger during one of his final autopsies.

And how can we not remember with anger the researchers who walked out of the laboratory in which they were working with Louis Pasteur, scandalized because he had introduced pathogenic agents into a nine-year-old child in what was, in fact, the first human vaccination against rabies?

New Dimensions

Interview with Lisa Randall

> If extra dimensions really explain phenomena in our world, then there are experimental implications of this fact for the future. Our ideas will lead us to see evidence for extra dimensions.
>
> —Lisa Randall

Eduardo Punset: Professor Randall, as I read your biography it reminded me of a famous professional trombonist—probably the first in the world—a lady whose name was Abby Conan, who applied for a job within the Munich Philharmonic Orchestra. No woman trombonist was in any orchestra in the world at that time. It so happened that a relative of someone on the try-out committee was also a candidate, so they put a

---

Lisa Randall is Professor of Physics at Harvard University.

screen at the tryouts to block the player from view. When Abby Conan began to play behind the screen, the orchestra leader, a world-famous man, cried out at once: "That's the musician I want! Let's stop the tests now!" Then, when they saw that she was a woman, a lot of noise ensued, several tried to block her appointment. The first years were very difficult for her, but later on they all lived happily ever after. Are you not one of the first great women physicists permanently employed as a faculty member in a place like Harvard University?

Lisa Randall: No, there are several other women. I don't want to say that I'm the very first. But I must say that anecdotes like yours actually circulated recently, in the context of the Harvard president Larry Summers's remarks last year, when there was much talk about hiring women faculty and accepting women students. This example of yours is very nice because it is so clear! There was no way to deny the evidence of bias, whereas more extenuating considerations come into the process of hiring. I remember when I played the flute in an orchestra as a young kid, a woman friend of mine did play the trombone. She thought it very funny to blast in my ear! Sorry, but this is what I remember about women who play trombones.

E.P. Today, while I waited downstairs reading your book *Warped Passages* in the snow, I enjoyed it so much, and it was snowing more heavily than it is now. An old Harvard professor came by, saw the book, and said: "Hers is a very good book, you know?" I would like our TV audience to enjoy your book as much as I enjoyed it.

L.R. Thank you.

E.P. And, if you don't mind, I will try to first explain to them the business of new dimensions.

L.R. Sure.

E.P. What we talk about in this conversation is new dimensions and how these ideas of new dimensions started. I will first remind everyone about "misinformation," as you call it, that the concept all started in the playpen. When we were small we could go forward and backward—we could go left and right and when we grew a little we even could go down and up, we jumped around. We should be very used to adding to three dimensions, but apparently, Lisa, we are so used to only them that it is difficult to imagine more dimensions, no? How is that?

L.R. Probably there's even more than that we're used to it. We really are physiologically designed to only picture or experience three dimensions. But we understand the physics of why there can be more. . . . If you think about the history of physics, certainly in the last century, we realize that things look very different than what we perceive if we detect them at either very small scales or very large scales! Who would have believed that quantum mechanics is the way the world works? Because we don't naturally see things on the atomic scale. Dimensions are like that: we experience three dimensions on the human scales of experience. Perhaps we are constrained by light, for example, which only does know about three dimensions.

E.P. One example I find most revealing is this one: when a toy is set before a camera with a backlit screen the toy is three-dimensional, of course, but its reflection on the screen is just flat. . . .

L.R. Yes, exactly, we have a single projection that doesn't have all the information. Your toy may have many different shapes yet in its two-dimensional projection, like a shadow, it may look exactly the same. But if you record all of its projections from every possible angle then you may re-create what you start with, even when limited to lower-dimensional information. That is what we try to do in physics: re-create a higher-dimensional world even though we only see the three-dimensional world.

E.P. Should we be so surprised about this? In our everyday lives, we try to reconstruct from images, from flat two dimensions to the real three-dimensional world.

L.R. Yes, it's funny. Just yesterday I spoke to someone about this. Even though in physics we constantly make amazing discoveries over centuries—admittedly, when we look over very different-length scales, with very different energies, things look completely different to what we see in daily life. Nevertheless, every time this happens it is shocking! [See table 2, chapter 33.]

E.P. We are shocked. So it would be if we were to discover extra dimensions, even though it is kind of natural. Why after all should there be only three dimensions just because we see them? There's no law of physics that demands only three dimensions. Einstein's equations, his general relativity that describes gravity, works for any number of dimensions. So why should there not be more?

You say, in your marvelous book, that physics explores different scales of matter or different or bigger lengths in the cosmos. You say that if there were further exploration in physics we might discover new dimensions. What makes you suggest that?

L.R. Yours is a good question. I think more dimensions might really be there and we might find them. Extra dimensions may or may not be there, but they may always be invisible. There are phenomena, mysterious things in our world, relationships among masses of particles that we see and understand their interaction. Some things exist that we, as particle physicists, have trouble understanding in the context of only three spatial dimensions. Some mathematical descriptions fit really nicely if we add an extra dimension in space. My co-worker and I found warped geometry, that is, different gravity in different places, very different scales in very different places. We found with warped space-time we could actually help explain some measurable phenomena. If extra dimensions do explain phenomena we see, then we have hopes of discovering them, say, within the next five years, with the large hadron collider (LHC) at CERN. . . .

E.P. In Switzerland?

L.R. In Switzerland, yes. What happens there is: They will accelerate protons to extremely high energies, and with such high energies, the collisions will create so much energy that heavy particles, particles with big mass (E = mc2) will be made such that these particles may give us fingerprints of extra dimensions. They may be particles that travel in extra dimensions, and have extra energy, which appears to us as extra mass because of the extra dimensions. The idea is if extra dimensions really explain phenomena in our world, we have a chance of discovering them.

E.P. You suggest that through the large hadron collider (LHC), in Switzerland, you may discover evidence of particles from other than the standard four dimensions.

L.R. It's amazing! It's true. Not that we will find all particles that travel in extra dimensions, but if they connect to real phenomena, we have already observed various masses and from this we infer that we have already measured the strength of gravity! If what we have already observed is connected to the weakness of gravity then we should be able to directly detect the other dimensions. It's so exciting! If extra dimensions really explain phenomena in our world, then there are experimental implications of this fact for the future. Our ideas will lead us to see evidence for extra dimensions!

E.P. I find fascinating an example of yours: you say the ideas of new dimensions have already, in other situations, helped us to understand mysterious things. You say, for instance, when you look at a world map, you see the continents as the tops of tectonic plates, and the way they are now and that they have moved around, of course. We only understand that, you say, because of the added extra dimension—time. If we think about millions of years in the past we begin to understand the way the world works. You imply the same idea for gravity. New dimensions might help us to understand this odd, mysterious, weak, omnipresent force of gravity. Gravitational forces are contrary to expectations. . . .

L.R. Yes. You're just competing against the entire Earth picking up that cup.

E.P. Incredible! A young physicist in Geneva actually, when I was working with *Redes* a few years ago, talked about the mystery of the weakness of gravity. Something we don't think about every day. A magnet can lift large weights against the ever-present force of gravity. And this young physicist said, "Listen, probably, you know, there might be other dimensions in which gravity is very strong, even though here, where force is used in our three dimensions, gravity is very weak." Do I make sense?

L.R. A number of possibilities exist. Our idea is that, actually, as one travels out in an extra dimension, gravity may change from extremely strong to very weak. In fact, we suspect that gravity varies exponentially, quickly. I mean, as you go out in the extra dimension, it varies. Gravity becomes exponentially weaker. Gravity goes down so quickly. So if you are any distance from where gravity is concentrated, as we say, you see gravity is extremely weak. Another dimension gives us a very natural explanation for why gravity is weak. This actually happens in the warped extradimensional framework that we concoct.

E.P. Another question on this business of gravity: you suggest we might be "trapped" in a world where really we have no access to other multi-universes. . . .

L.R. The suggestion, basically, is that we are in Flatland, except our flatland is three- rather than two-dimensional. So we live on a three-dimensional slice of a higher-dimensional world. Physicists invented a technical name, "branes." This comes from the word membrane in English. The idea is that we are in this three-dimensional flatland. The stuff of which we are composed, atoms and molecules, et cetera, our galaxies, our world, is really all stuck in a three-dimensional universe, this three-dimensional brane, even though extra dimensions exist. This doesn't mean we do not communicate with the external world, we do; because after all, gravity is experienced everywhere.

E.P. Everywhere.

L.R. Gravity connects to the entire geometry of space-time. What would it mean if we are stuck in a three-dimensional world? Well, it means, for example, that electromagnetism may only be experienced if one lives on this three-dimensional brane. So, for example, photons, that communicate electromagnetism, perhaps they are only on the brane! Electrons that are charged under electromagnetism, perhaps they are just on the brane. So maybe all the forces in particles we know about, except gravity, might too be stuck in our brane. There may be other forces in particles that exist on other branes elsewhere. Perhaps there is a multiverse; other universes may exist that we don't experience. But we are only connected, we know, to the outside world, via gravity. Might other forces connect us? Is this why we only experience three dimensions? After all, photons only travel on the brane. We can only see three dimensions. They just don't go out into the extra dimensions. Perhaps only gravity connects us to the extra dimensions.

E.P. Please elaborate a little on the idea of branes.

L.R. Sure. Again, we face the fact that we draw on a two-dimensional surface. So instead let's pretend branes have only two dimensions.

E.P. OK.

L.R. We think, in fact we live, of course, in three-dimensional branes, but let us imagine there are two dimensions. Then you could imagine that there's one brane here, the one we live on. Electrons and quarks, our atoms and galaxies we imagine, are all stuck on this one brane. It looks two-dimensional in this picture, but really it is three-dimensional, and so forth. It is infinite along the three dimensions. But let's suppose an extra dimension. This isn't a third one, this is really a fourth dimension of space; a fifth space-time dimension. Then imagine that there are other branes elsewhere that have other forces in particles. They likely have a totally different chemistry.

E.P. Gravity then would be the only force communicating from one brane to another?

L.R. There might be other forces we don't even yet know about but we do know that gravity would be able to communicate among them. So what happens? What you can imagine—let us enlarge one of these branes first. On this brane you imagine that you have here a proton that collides with another proton. The two protons collide. And when the two protons collide they change into gravitational particles, which are analogs of photons. The new particles communicate gravity. We call them "gravitons." No, we haven't seen them; you're right. We have not yet seen them. Actually, if our warped theory is right, it's interesting because of its experimental consequences. We would see evidence of extra dimensions. The graviton could actually be a graviton that travels in extra dimensions, which implies that we would see heavier particles.

E.P. What you suggest is that from math and physics, there may be a richer, a more varied, multidimensional different world, a cosmos beyond the one we are used to seeing. . . .

L.R. Some of these ideas are really eye-opening; it's almost as if the physics told us the ideas. We weren't really imaginative enough to think of them. We never thought there might be infinite extra dimensions we don't see. This conclusion came out of the math. It was a consequence of Einstein's equation of gravity. We looked carefully at a particular analysis and noticed what it meant. These are amazing ideas. We also find that we might be in what we call a "sink hole of extra dimensions." We might be in a region where we see three dimensions, even though elsewhere not three but many dimensions exist. The world is higher-dimensional. There is a variety of possibilities, and we do not know which is correct. This is frustrating for people who don't do physics, but it is frustrating for physicists, too. So many different possibilities.

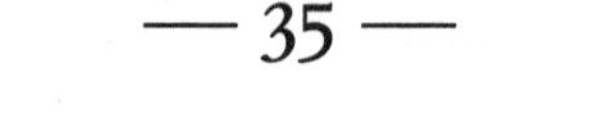

Manipulation by Dwarves

Interview with Nicolás García

> The human body has a terabyte of memory, ten to the twelfth power [10^{12}]. But if nanomemory were a thousand times larger—the technological development of planet Earth is very limited compared to the possibilities that exist.
>
> —Nicolás García

Nicolás García, director of the nanotechnology department of the Spanish Scientific Research Council, collaborated with Rohrer (chapter 32) for a long time. It was Nicolás who, with his tunneling microscope, showed me an atom for the first time. That atom sealed a friendship expressed in e-mails that is just as valuable as longer, more profound friendships in the nonvirtual world. He

---

Nicolás García, a physicist, works at the Center for Strategic and International Studies in Madrid.

never doubted the construction of the world from the top down, nor that machines will surpass the brain of the hominids when they have similar quantities of information. Such science-fiction fantasies are now part of science. They may come true before the end of the lives of those that imagined them. Very few scientists on planet Earth—and they can be counted on the fingers of one hand—maintain the same passion for their subject of those who at the outset of their research were at the forefront of nanotechnology.

Eduardo Punset: Nicolás, will we find robots in space if we find life beyond the Earth?

Nicolás García: Instead of seeking them, if they exist, they will visit us. If life exists elsewhere in the universe, it doesn't necessarily have to be like ours. Life uses soft materials. Maybe life-forms more complex and intelligent than ours are capable of traveling to greater distances in much less time. The technological development of planet Earth is very limited compared to the possibilities that exist.

E.P. You don't seem happy with what we have. Our bodies could be better, no?

N.G. Our bodies are the way they are, but larger, much more perfect beings, with much greater vision of the future, who understand nature better, could exist. Could life be out there? Of course. The search for extraterrestrials is complicated. It will take a long time to find them if they exist. If they are much more intelligent, they have to come to us. The galaxies are so large that our technology has no current ability to cross them. If life exists that moves much faster, with a much more advanced technology, it will come to us.

E.P. However, we already do wonders. When I see you, or I look at a woman, for example, what happens? If she is attractive . . .

N.G. I don't know about you, but when I see a woman it is demonstrated that in a second I take millions of photos of that woman and, naturally, the sexual aspect intervenes. I take these mental photos of the most seductive areas of that woman.

E.P. Millions of photos—

N.G. Your eye takes them, analyzes them, provides overall information, though you are not aware of all that information. You "photograph" only parts of her, but you have an overall vision of that woman in your consciousness. The visual system is not just a camera; rather it is more a computer that discards millions of bits of data and retains what is needed. Two different people have two different visions of the same object. The education of the person defines what they think they see.

E.P. For questions of data and memory, what is the problem? Instead of a referee in a football match who calls "out of bounds," why could not a machine call "out of bounds?" We have sufficient sensors to know what is happening. Nicolás, you said it is a question of sensors.

N.G. The invention needed is memory to register the images. If I had more memory, I could use part for interactive analysis of images and draw conclusions. We need to store much information in tiny spaces. We encountered similar problems in space travel. Could we have gone to the moon with old lamps and nineteenth-century technology? Information devices had to be miniaturized. The transistor was invented; that was a great revolution. The technologies need integration.

E.P. The next step is nanotechnology?

N.G. Of course. We are on the way. Silicon technology is at a quarter of a micron (0.00000025 m), but it could be reduced to atomic size. If we were capable of making an atom a bit of memory, we could have huge memories, of ten to the fifteenth power, a one followed by fifteen zeros, in a square centimeter.

E.P. With this memory, you say, I observe and store, and also I use part to make better decisions than my human body does now.

N.G. The human body has a terabyte of memory, ten to the twelfth power. But if nanomemory were a thousand times larger—

E.P. One could make better decisions—

N.G. This requires study in depth, but the data exists. The human body does act as a sensor that registers and analyzes data. That is my opinion.

E.P. Where will this lead to, say, in five hundred years?

N.G. In five hundred years, perhaps our devices will be more capable of decision making than we are. They will be autonomous. This certainly could happen; by cloning we could create such a perfect being that it would take us over. Not much we'll be able to do. I see this as a great problem but nothing will stop mankind's curiosity to create. If someone creates something superior, they will. Imagine that every human had a robot-computer added to him or her, able to communicate with others in real time and combine all the information! One, in addition, who was capable of making decisions. The results of such chaotic interaction are absolutely unpredictable, but not therefore senseless. Robots may make intelligent, successful decisions; it depends on their codes, on what they have learned. Exactly the same as what now happens with humans.

E.P. One tedious repeated claim is that in the end, a human mind always must program the robot.

N.G. In certain civilizations cutting off heads and human sacrifice were acceptable. Since then our codes have evolved. We have changed programs and behavior. We have become more humane because we think this is the best way to get on well together. But we too are programmed, like any other object that behaves!

E.P. If the nanorobots function and communicate with each other for quite some time, they will do things we will not know. We act in ways we haven't been taught, but that we discover through direct communication with the environment. New ideas arise and we evolve from interaction. When nanotechnological systems become sufficiently intelligent and interrelate, they too will evolve like humans.

N.G. More or less. It amuses me when people say that people don't work with a code. Of course we work with codes and they change naturally, with our interrelation: the interreaction of robot-computer programs with flexible code in continuous communication with each other must lead to the evolution of millions of interactions. Why not? It's obvious.

E.P. No state could stop it.

N.G. No state, or any moral consideration, either. It comes from curiosity! Humans have always been curious. Who convinced Columbus not to seek America? Nobody. The scientists of the time said that it was foolish. Who stopped the men who went to the Moon? Nobody. Such explorations can't be contained. "Whoever blinks first loses." And in science, curiosity is a universal truth.

E.P. Do you imagine you will be hounded, assailed by an intelligent machine in three hundred years?

N.G. Man, it's true, is a perfect animal, but no more than a monkey or another type of mammal, even though man is an animal who makes decisions (Margulis and Asikainen, 2008). Perhaps the most wonderful thing about a man or woman is that he or she is can write a beautiful novel. I don't know if robots will write novels like *Don Quixote*, *Madame Bovary*, or *Crime and Punishment*. But they will do other things with great capability, I think. Why not?

E.P. What you do is not virtual. You work with the essence of matter and life. You examine atoms. Physics Nobel Prize winner Richard Feynman said he didn't want slides or microtransparencies. He wanted to write and make things with atoms but he needed to see them. Until recently that was difficult.

N.G. In my village they say: "Something you can touch is closer than something you can see." Atoms were visualized some time ago. With the development of new technologies, the near-field microscopes, such as the tunnel-effect or the atomic-force microscopes, we have been able to touch atoms. We detect them by touching, like the use of Braille by a blind person.

E.P. How does that work?

N.G. A tunnel-effect microscope consists of a point that touches and describes atoms by moving parallel to the sample and detection of a current. The point moves at a distance of less than a nanometer (0.000000001 meter), which is amazing! One has to avoid vibrations so that the microscope has both stability and high resolution. Once achieved, you do see atoms. No system in nature is perfectly ordered throughout its entire extension. Some are ordered by parts. If they were totally ordered the system would be inert. The atoms move constantly due to temperatures above absolute zero.

E.P. Life requires imperfection. At what speed do the atoms move?

N.G. In the atmosphere, the atoms move at 1,500 meters a second. But in the case of this figure, they jump every nanosecond (0.000000000001 second). The atoms move on a surface in billionths of a second.

E.P. Is this really the essence of life?

N.G. The essence of life might be the possibility of finding places, defects, where atoms group together to form new structures. We would like those structures to be intelligent, nanorobots. Bill Joy interprets them as replacements of our future.

Space Destroyed and Time Obliterated

Interview with Paul Davies

> If you asked me whether in a hundred years time we will travel into the future, I would tell you that it is very possible.
>
> —PAUL DAVIES

A hundred years ago, no one believed that we could travel into space; now, however, we find no physical objection to the thought that in the future we will be able to travel in time.

Meanwhile, people try to understand what Einstein meant when he referred to time as an illusion, and learn to distinguish between physical time

---

Paul Davies is Professor of Natural Philosophy in the Australian Centre for Astrobiology at Macquarie University.

and psychological time; everyone has had the experience of feeling that the journey to the restaurant is longer than the journey back home, or enjoys breaking the barriers of space and time in dreams.

Many years before Einstein formulated the special theory of relativity, mankind had already sensed and experienced for itself that clocks and calendars do not reflect the real process of the passage of time. One of the most frequently repeated exclamations in the course of evolution by mammals with the capacity for language is "life is too short!" without thereby excluding those who, both from reason or due to schizophrenia, consider it to be too long and full of events. The fact is that mentally we are more prepared for traveling in time than for traveling in space. The Australian Paul Davies, an internationally recognized physicist, has dedicated much of his life to designing a time machine.

Paul Davies: We travel in time, to a certain extent, but only into the future, never back into the past. The difficulty in traveling into the past still depends on very advanced technology. If you asked me whether in a hundred years time we will travel into the future and to great distances, I would tell you that it is very possible, though I think that it will take much longer to find out how to travel into the past.

Eduardo Punset: Let's talk traveling into the future. I mean, it has already been demonstrated, with so-called atomic clocks, that time is not a universal, absolute value, as Newton thought. As you say, "my time is not your time," and perhaps the best way to start would be by recalling the story, or the paradox, of the twins.

P.D. I call it the twins effect, introduced by Einstein in his famous article from 1905, which formulated the so-called special theory of relativity. It represented a definitive change in the Newtonian idea that time is universal and absolute, identical for everyone everywhere. Einstein said that his time and my time can be different if we are moving in a different way. He illustrated it with the story of the twins. Imagine that we are twins and I travel in a rocket at a speed close to the speed of light for one year of my time. On my return, ten years will have passed on Earth. I will be convinced that a year has gone past and you that ten years have passed. Both of us would be correct and, in fact, I will have advanced nine years into your future, because ten years have gone by for you and just one for me. So if we did this in January 2005, I would only take one year to reach Earth year 2015. It is, in a way, a journey into the future, a real effect. I have described it in terms of advancing years, though it can't yet be done because we can't travel at a speed close to the speed of light. At the most, we can travel at 0.1 percent of the speed of light. And at those kinds of speeds the effect is very small, but real, and it is possible to measure it using atomic clocks. It is not therefore a theoretical prediction, but rather a real effect. Of course, it is possible to travel in time and the faster you travel the further ahead in time you go.

E.P. Paul, correct me if I'm wrong, but you mean that if we could travel at a speed close to that of light, we could travel to a star that is at a distance of one hundred light years in ten years, a very human period of time, right?

P.D. That's right. One can travel very long distances in relatively little time using the so-called time warping or time dilation effect. All we need is to observe the objects that travel at the fastest speeds, such as cosmic rays that travel almost at the speed of light. High-energy cosmic rays can cross the entire galaxy in minutes—measured in their reference system—but to us, observing from Earth, they appear to take one hundred thousand years to cross the galaxy. So, by means of the time warping effect at high speed, a journey of one hundred thousand years would only take, in reality, a few minutes. That is just one example of the enormous effect of time warping. I don't think that even the writers of science fiction have imagined that we could have space ships that could travel at speeds close to that of light. But it is feasible to imagine a new propulsion system that could propel a spaceship to 90 percent of the speed of light. Then these time-warping effects would be very significant.

E.P. comments: While Paul was speaking, I thought I better understood the crumpled image of space-time in which a celestial body gravitates. He was simply talking about speed, thinking of the effects of special relativity that introduced the concept of relative time. Time, to which Newton referred as an absolute magnitude, in fact depends on movement, so time passes more slowly for the observer traveling at a greater speed. Gravity apparently causes this contraction and dilation of time, as well. According to the most revolutionary postulate of general relativity, the space in which we live is curved and gravitation is the manifestation of this curvature. Therefore, the movements of the stars can be understood as a simple fall of the lighter bodies toward the heavier ones, due to the deformation of space around the objects of greater mass, like the stars or black holes.

E.P. It's incredible. Isn't it very confusing given our old way of measuring time? We differentiate the present, the past, and the future as if time flowed regularly, but from what you say it is possible that time does not flow from the past to the present to the future. Is it an idea we have to abandon? I mean, will scientists bring an end to the flow of time and perhaps the anxiety caused by death or birth?

P.D. Einstein wrote a letter to a friend in which he said that distinction of the past, the present, and the future is just an illusion, although a very persistent one. He said that we physicists consider the past, the present, and the future as constructs of the human mind that, in reality, do not exist. And the present and the future can be questioned as a result of the effects of time warping. It is an important message of Einstein's theory of relativity: if I can travel in a spaceship and I return to Earth and I realize that the two twins are no longer twins, then questions like "What is my twin doing now?" will make no sense. If I travel very fast in a rocket, my time and your time are no longer synchronized and there is no shared "now" or present. And this argument also serves for the concept of being in the same time. Einstein formulated it in his article in 1905, and since then physicists have accepted the conception of time in its totality, instead of divided into the present, past, and future. I like to use terms like *timescape* or *landscape*, which are quite widespread. Physicists think as if all time had been established since before human existence, without present, past, or future. In human life, though, it makes no sense, because it is necessary to be able to say here or now. But if we ask what is happening on Mars now, the question is ambiguous, because the event on Mars to which we refer depends on the observer's reference system. And with greater distances, if we ask what is happening in a very distant galaxy now, then the definition of what "now" is might vary by thousands of years, depending on whether we are sitting or walking, owing to the effect of time warping. So, this clear division between the past, the present, and the future disappears if we take into account the effects of the movement of the observer in time.

E.P. In the case of Mars, the difference is twenty minutes, right? It's the time that information takes to arrive, so saying "now" is very ambiguous if that "now" refers to Mars.

P.D. There is a range of moments and instants of time that could be considered at the same time as our conversation.

E.P. The most complex question does not refer to traveling into the future, but rather into the past. In your books, you say that it may be absurd if certain paradoxes aren't resolved. A paradox reveals the absolute impossibility of traveling into the past. If a person travels into the past, he meets his grandmother and kills her, the paradox appears because the killer could not have been born. How can this paradox be resolved?

P.D. First of all, it must be said that travel backward in time, into the past, is much more problematical than travel into the future. It's irrefutable that we cannot travel into the past simply by the process of traveling at a speed close to that of light, because in that way time only moves forward, not backward. To go back in time, a much more complex method is necessary. My preferred method is the black hole in space. To answer your question about whether we can travel into the past and what prevents us, one answer is that if we change the past it might not be coherent with the future. The most famous paradox is the one you mentioned, very well known by lovers of science fiction fascinated by tales about time travel. For example, if we change the past by killing our own grandmother, the question arises of how the grandson could have been born. And if he wasn't born, how could he travel in time and commit the crime? Science gives a coherent description of reality. It would not be scientific if there is a true paradox. But this does not mean that we can't be part of the past or that the causes can't go back in time. I'll give you a very simple example: imagine that instead of traveling in time and killing his grandmother, the traveler goes back in time and saves a young girl who is drowning in a river, and it turns out that she is his grandmother. Then, history is coherent. So, as long as the causal and temporal relations are coherent, there is no fundamental objection to traveling into the past or sending signals into the past. But the causal relations must be coherent; they can't be incoherent paradoxes. However, this does not mean that we can't say it is impossible to travel into the past.

E.P. So, let us analyze how we could make that journey, if it is possible. But, first of all, let's try to resolve another paradox that is often given as an example of the impossibility of travel into the past. There are no time tourists, people who travel from the year 3000 to the crucifixion, in the year zero or 33. But if time travel did exist, many people from the future would travel into the past. What do you think of this paradox?

P.D. You're right, if time travel were possible, a traveler from the future would come to this study and demonstrate to us that it is possible to travel into the past. Therefore, there are no tourists from the future. But from our current understanding of physics, all the time machines that have so far been suggested cannot be used to go back in time to a date before the construction of the machine. So, if in a hundred years time we manage to discover a time machine that travels into the past, we will not be able to use it to return to this present, because it takes more than one hundred years to create a temporal relation of one hundred years. Therefore, for example, it could take one hundred and ten years to create a machine that could go back one hundred years, but it can't be used to go back two hundred years. There is no possibility that our descendants will build a time machine with which they can visit us. The only possibility of meeting tourists from the future is if an alien civilization had made a time machine long before and they had lent it to our descendants. Then they could travel to our present. So the fact that time tourists do not appear only means that it hasn't been possible to build the time machine or that there are no aliens that have lent one to our descendants.

E.P. We conclude that we had better hurry and construct the time machine, because the people of the future will only be able to go back to the time when the machine was built. In your wonderful book *How to Build a Time Machine*, three absolutely essential things to build a time machine are listed: a collider, an imploder, and a differentiator. Might you please explain that a little?

P.D. The essential aspect of the time machine that I suggest is called a wormhole in space. I'll try to explain what that means, as many people are not familiar with the concept of black holes. You can enter a black hole, but you cannot come out; it's like a one-way journey to nowhere. A wormhole would be like a black hole, but with an exit as well as an entrance: you could look in one end and see the other end. If there was a wormhole in my study in Sydney and I jumped into it, I would not come out in some other part of the city but rather on the other side of the galaxy. Sometimes, in science fiction, the wormhole is called a star gate. Carl Sagan popularized it in his novel *Contact*, in which a wormhole appears with an entrance in Japan and the exit near the star Vega. If you have seen the film based on the novel, you will remember that Jodie Foster is thrown into this wormhole, inside a spherical object. At the opening, the wormhole looks like a gigantic liquidizer. Science fiction has often used the idea of a star gate or a wormhole that connects two distant points in space by means of a short cut, so you can jump in and travel great distances. This idea, which was born in science fiction, was studied by a group of scientists from Caltech, the California Institute of Technology, who demonstrated that if a wormhole did exist in space, it would be possible to build a time machine. It would be adapted in such a way that if you jumped into the hole not only would you come out somewhere else in space, but also in a different time, past or future, depending on how you jumped.

Therefore, the first step in my proposal to construct a time machine is to find a wormhole. But that's not easy. In space, either wormholes don't exist or we have not been able to find them yet. What's more, if they do exist, they will be very small, about ten to the minus twenty (10-20) of the size of an atomic nucleus. An atom will be gigantic compared to a wormhole, but certainly it wouldn't be big enough for a human being! Because we know virtually nothing about the physics of wormholes we face a very difficult task. We have to find the way to reach the little interstices and take control of a wormhole and enlarge it.

The first step will be to use a particle collider, enormous accelerators that make atomic subparticles revolve at speeds close to that of light and make them collide. If we could do this with a great deal of energy, we could create small wormholes. In Switzerland, a large collider is being built, at the European Laboratory for Particle Physics, known as the European Nuclear Research Center. It has been called the large hadron collider (LHC), and it will be ready in five or six years. If certain exotic theories on gravitation are correct, microscopic wormholes could be created in this machine. We can't use them as time machines as they will be smaller than an atom, but they will demonstrate that wormholes exist. To travel in time, we will have to expand a wormhole to human size. We know that, to prevent wormholes from collapsing and turning into black holes, negative energy must be injected into them, an idea about which there is much speculation. Though it seems like science fiction, negative energy can exist and be created in the laboratory by lasers of a certain configuration. Small packets of negative energy form. An enormous amount of negative energy will be necessary to expand the wormhole. We still don't know how to do that. I speculate in my book on this difficulty. Therefore, if we combine these two technologies—the collision of subatomic particles to create the little wormholes, and then make them bigger with negative energy—we will have taken the first two steps towards building a time machine.

E.P. Paul, might nature spontaneously create a wormhole? Could a wormhole appear in the future?

P.D. It's possible. As I said, a black hole has negative energy inside it. How sufficient negative energy can be supplied to the wormhole to make it grow to human size is difficult to imagine. But gravitational fields of the wormhole itself might create exotic quantum effects, related to the subatomic nature of matter. The wormhole might even create its own energy. The wormhole might even expand by itself. Then, another difficulty: preventing the wormhole from becoming too big. This exotic process may have already happened in nature, for example, just after the big bang, when the universe was created. The exotic quantum effects may have created wormholes in space, but we haven't found them. Astronomers look for signs of wormholes, though it's difficult to imagine how they will manifest themselves. Perhaps, by use of certain kinds of astronomical examination, we might find wormholes in the same way that we found black holes to exist. If wormholes existed, we wouldn't have to carry out the complex operations I've described in order to obtain them from quantum space-time inside an atomic nucleus.

E.P. A previous interview examined the mystery of how and when inert matter gained life [see chapter 28]. You astronomers look for signs, "biosignatures" of complex processes such as life on Mars [see chapter 24]. Paul, you suggest the possibility that life began on Mars, not on Earth. Your idea is that a living organism landed on Earth inside a rock launched by a meteorite from Mars. But why do you think that it was a rock from Mars that brought bacteria to Earth and not the other way round?

P.D. We don't know the chemical processes that gave rise to life. However, we do know that life appeared first on Mars and then it came to Earth.* Why? Because Mars was ready for life long before Earth, about 4.5 billion years ago. As a smaller planet, Mars cooled down more rapidly. During this time, Earth was still too hot, and the planets were bombarded by enormous asteroids and comets. Their surfaces were dangerous places. I think that life began under the surface, two or three kilometers beneath the ground, when the surface above was still very hot, until about 3.8 billion years ago. Therefore, I think that life arose on Mars because it cooled down more rapidly. In any case, in the course of history many rocks from Earth have traveled to Mars and have carried life in that direction. But I think that life originated on Mars and came to Earth inside those rocks that were broken off the red planet by comets and asteroids that crashed into it.

E.P. And, as Mars is a smaller planet, the force of gravity is less and it was easier for rocks to be launched into space after being hit by meteorites, right?

P.D. Correct. More traffic of materials occurs from Mars to Earth than in the

*No direct evidence exists either that life began on Mars or that it was transported by meteorites to Earth, in spite of this confident statement. —LM

opposite direction because Earth's mass is greater. It has been easier for rocks to go from Mars to Earth.

E.P. One last question has appeared time and time again in the course of our conversation: the relation between life and the processing of quantum information? Quanta and time travel? Can you explain it so that we understand relations between time travel, the processing of quantum information, and life?

P.D. The subject that links all these questions is "information." I think that life does not appear by the mixture of the right chemical products in the right way.* A living cell is not merely a magic formula, it is a system of information processing and response. I think that the secret of life lies in understanding information. Our concept of the nature of information has been transformed owing to the development of quantum computing in the past ten years. Scientists try to store and process quantum-level information by use of the strange rules of quantum physics, as we discussed. The same strange rules of quantum physics that apply to the production of the quantum wormhole and the idea of time travel use many of those ideas. The paradoxes that appear in time travel are also related to information. The problem is that by sending information into the past these paradoxes are generated.

Another idea of science is appearing, in which information instead of matter plays a leading role. Physicists, for centuries, have thought matter to be the basis of physical theories. Now a movement is afoot that considers information as primary and matter to be what arises from a correct understanding of information. It's simply conjecture that might appear rather far-fetched, but I'm not the only one who has suggested it. John Archibald Wheeler, the physicist who invented the terms *black hole* and *wormhole*, also believes that information is the basis of everything. Wheeler expresses it in a wonderful, very poetic way: he says "it from bit." He refers to bits of information when 1,000 bits make 1 byte. He suspects that physical objects will emerge from bits of information. This process of linkage between information and matter is suspected, though now we can only speculate because we don't yet fully understand.

*William Day agrees, as he explains in chapter 28.—LM

THE SCIENTISTS

I. People Primates

Culture before humans

1. Nicholas Mackintosh
Mackintosh, for many years since 1981 and until his retirement in 2002, worked at Cambridge University in England as a Professor of Experimental Psychology. He had received both his undergraduate and doctorate degrees from the oldest, and many would say the most distinguished, university in the British Isles, if not the world: Oxford University. In his long and impressive academic career he has held posts at Dalhousie University in Halifax, Nova Scotia, Canada; Sussex University in the south of England; University of California, Berkeley; University of Hawaii; Yale University in New Haven, Connecticut; New South Wales University in Australia; University of Paris-Sud in France; and in Pennsylvania at Bryn Mawr College.

2. Robert Sapolsky
Robert Sapolsky, who currently teaches at Stanford University in California, is a superstar. A recipient of a "genius award" as a John D. and Catherine T. MacArthur fellow, he has also been lauded with Stanford's Bing Award for Teaching Excellence. His full title is John A. and Cynthia Fry Gunn Professor of Biological Sciences and Professor of Neurology and Neurological Sciences at Stanford Medical School.

As an undergraduate at Harvard University in Cambridge Massachusetts he received his bachelor's degree in anthropology with honors. His graduate work at Rockefeller University in New York City led to his doctorate for which he did original research in endocrinology. He has always been interested in stress exacerbated by social relations but not exclusively in people, as is clear from his popular 1994 book *Why Zebras Don't Get Ulcers*. Much of his work has been in Kenya with baboons in the wild (they are large monkeys) and with chimpanzees. He studies, among many other behaviors and physiologies, endogenous steroids. He has, for example, worked on cortisol endocrine levels in the brain and their correlation with social hierarchy in animal aristocrats, the "alpha" males and their mates.

3. Jane Goodall

Jane Goodall, arguably, is the most famous woman scientist in the world but she did not begin life as a science nerd. Early in her forty-five year study of chimpanzees in the wild, primarily in Gombe Stream National Park in Tanzania, east Africa, she developed remarkable personal relationships with these fascinating primates and gave them names. She discovered the production and use of termite-sticks by chimps as tools for prying and coaxing these delicious, nutritious, social insects from their nests. She has described primate aggression, for example the co-operative hunting by chimp troops of red colubus monkeys.

Her honors are as extraordinary as her accomplishments. They have been conferred with regular frequency since 1980 when she was appointed an awardee of The Golden Ark by associates of the World Wildlife for Conservation. In 2004 Queen Elizabeth II named her Dame Commander of the British Empire in a ceremony held in Buckingham Palace. Two years earlier Secretary-General Kofi Annan named Dr. Goodall "a United Nations Messenger for Peace." Her work on behalf of our extant primate relatives through protection of their habitat is tireless. She is surely in tune with the message of this book when she classifies both species of chimp (*Pan troglydites* and *Pan bonobo*) with gorillas and ourselves in the hominid family.

4. Jordi Sabater Pi

Jordi Sabater Pi, now emeritus professor of psychobiology and ethology at the University of Barcelona, pioneered field research studies in western Africa. He spent more than thirty years with chimpanzees, gorillas, and even amphibians (the goliath frog) and birds (the honey guide) and people (the Fang) in natural habitats in equatorial Guinea. Between 1940 and 1969 he studied nature and documented his research in drawings, water colors, and scientific papers that are now on display in Barcelona. He deplores destruction of the African tropical rainforest and notes, in the areas he knows well, the jungle has lost 80 percent of its arial cover to human economic exploitation.

5. Edward O. Wilson

Wilson, now emeritus from Harvard University in Cambridge, Massachusetts, received his Ph.D. also from Harvard. He was born in Birmingham, Alabama, where he stayed at the University of Alabama for both his bachelor's and master's degrees of science in biology. A lover of termites, and even more so of ants, he is a gentleman of conscience and endowed with extraordinary biophilia—he has inspired many of us to express our love of other-than-human life. His insightful and evocative writing, much on insect and other societies, has led to the foundation of entire fields of science: studies of sociobiology and biodiversity.

Erudite, articulate, passionate, and yet humble in the face of nature and its beauties, Wilson is an extraordinary field naturalist, laboratory scientist, and conservationist. A prolific writer, he is recipient of two Pulitzer Prizes (one for *The Ants* with his former student and colleague from Wurtzberg, Germany, Bert Hölldobler, and one for the single-authored *On Human Nature*). He held a Crafoord Prize, the highest possible award in ecology, and since 1967 has annually published books and papers of wide significance. His current formal title, Pellegrino Research Professor in Entomology, Department of Organismic and Evolutionary Biology, fails to display his decades of superb teaching of evolution and ecology to crowds of undergraduates who are more than grateful to have been his students.

Attractiveness

6. Victor Johnston

Victor Johnston is now Professor Emeritus at New Mexico State University, Las Cruces, in the Department of Psychology. His specialized and original studies have been classified into new fields of science: "cognitive engineering" and "biopsychology." He has pioneered the search for the origins of human emotions, human perceptions of beauty, and genetic bases of body and facial stimuli, without transgressing beyond the confines of scientific research. He is the author of *Why We Feel: The Science of Human Emotions*. He continues with investigations that led to this book. The twenty-years-later relationship between *in utero* hormone dosage and adult mating preferences continues to fascinate him. He is confident that the correlation is strong.

7. Daniel Gilbert

Daniel Gilbert is Professor of Psychology at Harvard University and director of the Social Cognition and Emotion Lab. He has focused on studies of affective forecasting—the ability, or lack thereof, of people to accurately predict their emotional reaction to future events. We rarely find happiness where and how we think we will, so it seems. Though his research is serious science and he plays by the rules as he seeks evidence in examination of the inherent biases of human cognition, he is not prevented from communicating his ideas in entertaining ways. His essay on the failure of our society to take climate change seriously was called "If Only Gay Sex Caused Global Warming."

8. Robert Hare

Robert Hare, Emeritus Professor of Psychology at the University of British Columbia, has taught and conducted research for some thirty-five years. He is president of Darkstone Research Group Ltd., a forensic research and consulting firm. Most of his academic career has been devoted to the investigation of psychopathy and its nature. His assessments of the implications of psychopathic behavior for mental health and criminal justice are astounding to most of the rest of us. He is the author of several books: *Without Conscience: The Disturbing World of the Psychopaths Among Us* and *Snakes in Suits: When Psychopaths Go to Work*. He has written more than one hundred scientific articles on psychopathy.

Anxiety

9. Daniel Dennett

Daniel Dennett, Professor of Philosophy, directs the Center of Cognitive Studies at Tufts University, Medford, Massachusetts. A leading philosopher concerned with research and practice of artificial intelligence (AI), he maintains that the most important aspects of consciousness can be computed. He is a natural teacher, a warm and passionate intellectual leader. Dennett has been honored with numerous awards, and has been prominent in public debates on issues of evolution, consciousness, and—with his recent book *Breaking the Spell: Religion as a Natural Phenomenon*—the function of religion in society.

10. Oliver Sacks

Oliver Sacks was catapulted into fame by his charming and accessible writings. He described the psychological experience of his patients and others in *The Man Who Mistook His Wife for a Hat* and *Awakenings*. These bestsellers have inspired films and at least one opera. As clinical Professor of Neurology at the Albert Einstein College of Medicine, Adjunct Professor of Neurology at the New York University School of Medicine, and Consultant Neurologist to the

Little Sisters of the Poor, he is a well-known practicing psychologist and teacher. The depth of his chemistry education is far less appreciated but clear in his book *Uncle Tungsten*. He maintains a private practice in New York City. He emphasizes the perceptions and experiences of the people with whom he works rather than clinical detail in his writings. His masterful sympathy and verbal skills provide the basis for his immense popularity.

11. Rodolfo Llinás

Born in Bogotá, Colombia, Rodolfo Llinás currently holds the Thomas and Suzanne Murphy Professor of Neuroscience and chair of the Department of Physiology & Neuroscience at the New York University School of Medicine. His research ranges from the molecular level in neurons through to the cognitive level in an attempt to discern how consciousness and a sense of "self" emerge from gray matter. His book *I of the Vortex* is highly praised as an explanation of his theory of the mind. A lively, curious intellectual, he often can be found immersed in neurobiological research at the Marine Biological Laboratory at Woods Hole, Massachusetts, during his summer "vacation."

12. Joseph Ledoux

Also at New York University is Joseph Ledoux, the Henry and Lucy Moses Professor of Science, and Professor of Neuroscience and Psychology. Ledoux directs a multi-university research center: the Center for the Neuroscience of Fear and Anxiety. As the name hints, Ledoux primarily works on understanding how the mind-brain responds to information perceived to represent danger. His discoveries on connection between memories and emotions are detailed in his book *The Emotional Brain*. In *Synaptic Self* he broadens his view to the emergence of selfhood, which he locates in the pattern of neural synapses.

13. Kenneth Kendler

Kenneth Kendler studies the genetics of psychiatric disorders and substance abuse. As director of the Virginia Institute for Psychiatric and Behavioral Genetics at Virginia Commonwealth University, he is Professor of Human Genetics. Simultaneously he is the Rachel Brown Banks Distinguished Professor of Psychiatry. He has received scores of honors for his work. Kendler also happens to be a rather good banjo player.

II. Animal Body-Mind

Cyclicity and Sociality

14. Steven Strogatz

Have you ever played the game Six Degrees of Kevin Bacon? Someone names any actor and your task is to link him or her to Kevin Bacon through six or fewer intermediary actors who have appeared in the same movies. This pop culture example of "small world networks" is one of the themes that mathematician Steven Strogatz, Cornell University's Professor of Theoretical and Applied Mechanics, emphasizes in his research. Always alert to intellectual opportunity, stimulated by new curiosities, Strogatz contributes to fields others consider totally unrelated: physics, engineering, biology, superconducting Josephson junctions, DNA coiling, and the synchronicity of cricket chirping. Strogatz deserves his reputation as a true original.

15. Richard Gregory

How trustworthy are our perceptions? Hardly at all. Dr. Gregory has been at the forefront of studies of perception. He frequently uses optical illusions to understand how we perceive, and often misperceive, the world around us. His interest in perception is part of his larger interest in the general workings of the mind. These scientific passions led him to co-found the Department of Artificial Intelligence (originally named the Department of Machine Intelligence and Perception). He spent many years in Scotland at the University of Edinburgh. He is now Emeritus Professor of Neuropsychology at the University of Bristol.

16. Nicholas Humphrey

Nicholas Humphrey follows the path less taken. He followed early studies at the side of Diane Fossey and Richard Leakey with a career investigating the evolution of consciousness. Out of this he has written several popular books, *Consciousness Regained*, *A History of the Mind*, *The Mind Made Flesh*, and, most recently, *Seeing Red*. This latest is a fine example of Humphrey's willingness to challenge his peers with radically original thinking, for example, by arguing that perception and sensation are entirely distinct events and processes occurring in our brains. In addition, he has long been actively involved in the antinuclear movement, including co-editing the book *In a Dark Time*, for which he was awarded the Martin Luther King Memorial Prize in 1984. He is currently

School Professor at the Centre for Philosophy of Natural and Social Sciences, London School of Economics.

17. Diana Deutsch

As Professor of Psychology at the University of California, San Diego, Diana Deutsch focuses her research on the ways the mind handles auditory information, particularly music. She is best known for her marvelous discoveries of musical and auditory illusions. Her recent work delves into the concept of musical grammars, for example the ability to establish memories of musical pitch. The results of this research indicate that the capacity for perfect pitch is common, but early development of language skills as an infant and child determine whether the ability is established for life.

Prehistory and Immortality of the Body-Mind

18. Sydney Brenner

Sydney Brenner is a biologist whose efforts to understand messenger RNA and the process of genetic mutation were seminal contributions to the establishment of the field of molecular biology. His research into the workings of the nervous system of the nematode *C. elegans* earned him a Nobel Prize in Physiology or Medicine in 2002. Brenner was born in South Africa, where he remained until going to Oxford University to study for a Ph.D. While there, he befriended Francis Crick and he would later share an office with Crick for many years at Cambridge. In the 1990s he began working part time in the United States at the Scripps Research Institute in La Jolla, California. Later he established The Molecular Sciences Institute, and after retiring from there joined the Salk Institute as a Distinguished Professor—where he once again joined his old friend Francis Crick.

19. William Haseltine

William Haseltine is President of Haseltine Associates Ltd. and President of the William A. Haseltine Foundation for Medical Sciences and the Arts. He is also a professor at The Scripps Research Institute and sits on the board of the Institute for One World Health. In 1992 he founded Human Genome Sciences, serving as its chairman and CEO until October 2004. A Harvard University faculty member from 1976 to 1993, Haseltine received a Ph.D. in biophysics from Harvard University.

20. Phillip Tobias and Ralph Holloway

Phillip Tobias is a paleoanthropologist who, with Louis Leakey, discovered the *Homo habilis* species of pre-human hominid. He has published hundreds of scientific papers and numerous books, has been awarded many high honors, and holds professorships at the University of the Witwatersrand and Cornell University. In addition to his scientific pursuits, he was an ardent opponent of the apartheid system in his home of South Africa.

Ralph Holloway is Professor of Anthropology at Columbia University. His brain endocasts of hominids help us understand, from tissues that do not fossilize, how human brains evolved and provide direct evidence of early ape brain evolution.

21. Douglas Wallace

Douglas Wallace, the Donald Bren Professor of Molecular Genetics and director of the Center for Molecular and Mitochondrial Medicine and Genetics at the University of California, Irvine, studies human mitochondrial DNA and identification of mitochondrial mutations that cause degenerative diseases. Mitochondrial genes pass nearly unchanged from mother to daughter (and to son; there is no transmission of mitochondrial genes by fathers) and thus may indicate paths of ancient human migration.

22. Tom Kirkwood

Tom Kirkwood is Professor of Medicine, Co-Director of the Institute for Ageing and Health at the University of Newcastle, and Director of the Centre for Integrated Systems Biology of Ageing and Nutrition. Educated at Cambridge and Oxford, his research focuses on the science of aging and on understanding how genes as well as nongenetic factors, such as nutrition, influence longevity and health in old age.

III. Life on an Animate Planet

Bygone Biospheres

23. James E. Lovelock

James Lovelock was working for NASA as part of the Viking program to study Mars, when he realized that one strong characteristic of life on Earth is that living organisms alter the chemistry of the atmosphere and their biological activities maintain the atmosphere in a state far removed from what it would be without the presence of life. This lead him to develop an approach to understanding the biosphere of the planet, from near-surface geology up through the atmosphere, as an integrated system of living and nonliving elements best viewed as a singular whole. The result is "Gaia Theory," which has blossomed as a field of research initially dismissed by mainstream scientists and now—perhaps grudgingly—granted some modicum of respect as a legitimate topic of study. Its acceptance outside academia has been much stronger, as many people struggling with worldviews that seem unable to coexist with threats of environmental catastrophe take Gaia Theory to be a more sensible framework, often influencing their spiritual beliefs as well as their logical notions of how the world works. Lovelock continues to pursue his research independent of established institutions, working out of a converted barn in Cornwall. His popular books on Gaia Theory and its implications include *Gaia: A New Look at Life on Earth* and, most recently, *The Revenge of Gaia: Earth's Climate Crisis and the Fate of Humanity.*

24. Kenneth H. Nealson

Kenneth Nealson, the Wrigley Chair in Environmental Studies and Professor of Earth Sciences and Biological Sciences at the University of Southern California, has pioneered the field of modern geobiology—the still-young branch of science investigating where the processes and chemistry of life intersect with the planet's mineral and metal chemistry. In his early work as a marine microbiologist, Nealson discovered quorum sensing, the phenomenon in which microbial communities create light. On a much larger scale, he has studied the cycling of such minerals as iron and manganese, revealing the key role of microorganisms in these biogeochemical processes. More recently, he has turned to figuring out how life can function in extreme environments, and he is directing efforts at NASA's Jet Propulsion Laboratory to search for life and evidence of ancient life in the solar system.

Toward Perfection

25. Stephen Jay Gould

Stephen Jay Gould, who died in 2002, is one of the best-known scientists outside of scientific circles, largely because of his marvelous popular writings on a wide variety of topics from evolutionary theory to baseball. His numerous books include *The Mismeasure of Man*, which he updated in 1996 to refute the biased racial claims in the then-bestselling *The Bell Curve* by Richard Herrnstein and Charles Murray. This was just one example of Gould's dedication to an inclusive, humanist ethic, inseparable from his dedication to good scientific research and intellectual honesty.

He is probably best known for his development, with Niles Eldredge, of the theory of punctuated equilibrium, according to which evolutionary developments in the history of life on Earth occur in rapid spurts separated by lengthy periods of relatively little change. This theory stood in opposition to the simpler, traditional concept of random mutation and Darwinian selection, but the fossil record clearly supports the newer approach.

Within the field of biology, Gould is also frequently cited for a paper he wrote with Richard Lewontin arguing that many in the biological sciences confuse causes and effects when they dogmatically apply a theory of evolutionary adaptation to explain each and every trait in an organism. Instead, Gould and Lewontin argue, many traits are like architectural spandrels—they are the side effects of other developments within the organism. Having occurred, they then undergo adaptation toward usefulness, but this is within the context of the constraints of the other developments that created them in the first place. As they suggested, "Male tyrannosaurus may have used their diminutive front legs to titillate female partners, but this will not explain why they got so small." When scientists start their hypothesizing process with the smallness of the tyrannosaurus' front legs, and not with the whole creature in mind, they risk a train of thought that makes the small legs inevitable, which is really no explanation at all.

26. Richard Dawkins

Many readers have probably encountered Richard Dawkins's ideas without knowing it. His 1976 book *The Selfish Gene* led to the "selfish gene" phrase entering the popular lexicon, and it has become a classic text for the wing of evolutionary theory that emphasizes a reductionist approach. In the book he used the term *meme* to refer to an idea that moves through culture in a process analogous to the spread of a gene in a biological population. Others, seizing on the idea and the term, have developed an entire field of study called memetics.

Dawkins is a professor at Oxford University where he holds the Charles Simonyi Chair for the Public Understanding of Science. In addition to his work in evolutionary theory, he has been an ardent proponent of atheism and critic of creationism, as expressed in his latest book *The God Delusion*.

27. Dorion Sagan

The son of Lynn Margulis and Carl Sagan, Dorion Sagan studied history in college, but he has focused on scientific topics in his prolific writing career, frequently co-authoring with Margulis. His numerous books include *Acquiring Genomes: A Theory of the Origin of Species*; *Microcosmos: Four Billion Years of Microbial Evolution*; *Origins of Sex*; *What is Life?*; *Up From Dragons: The Evolution of Human Intelligence* (with John Skayles); *Into the Cool: Energy Flow, Thermodynamics, and Life* (with Eric D. Schneider); *Biospheres: Metamorphosis of Planet Earth*; *Dazzle Gradually: Reflections on the Nature of Nature*; and *Notes From the Holocene: A Brief History of the Future*. Sagan is also an adept sleight-of-hand magician.

Dead or Alive?

28. William Day

A native of Indiana, William Day is a graduate of Indiana University and McGill University in Montreal where he received his Ph.D. in chemistry. His research and writings have dealt principally with structural order and origins—in particular, he is fascinated by the "origin of life" problem. We have a reasonable understanding of how life produces more and new life, but how the very first bit of living chemistry emerged from the nonliving remains a sticky question. Day argues that life is not a mixture of chemicals in an oily bag but rather, from the beginning, the tendency to imperative, prodigious and continuous overgrowth.

29. Ricardo Guerrero

Ricardo Guerrero is a Full Professor of Microbiology at the University of Barcelona and also an adjunct professor in the Department of Microbiology at the University of Massachusetts at Amherst. His expertise is in the origin and evolution of the Earth's earliest ecosystems.

30. John Bonner

John Bonner, the George Moffett Professor of Biology Emeritus, joined the Princeton faculty in 1947 and served as chair of Princeton's Department of Biology before his retirement in 1990. He is a member of the American Philosophical Society and the U.S. National Academy of Sciences and a fellow of the American Academy of Arts and Sciences and the American Association for the Advancement of Science. His research has focused on slime molds, in an effort to better understand evolution and development. The slime molds are fascinating organisms that spend much of their life cycle living as separate single-celled amoebae, but under certain conditions join together into a multicellular fruiting body to reproduce through spores.

IV. Toward the Invisible

From the Vast to the Miniscule

31. Eugene Chudnovsky

Eugene Chudnovsky, Distinguished Professor in the Department of Physics and Astronomy at the Herbert H. Lehman College, in Bronx, New York, researches quantum influences in superconductors and the "tunnel effect." His 150 research articles concern mainly magnetism and superconductivity. He has long been concerned about the rights and freedoms of scientists in the international community.

32. Heinrich Rohrer

Heinrich Rohrer won a Nobel Prize in Physics (1986) for his contribution to the design of the scanning tunneling microscope, which allows visualization of subatomic scale detail. He works at the IBM Zurich Research Laboratory in his native Switzerland.

33. Sheldon Lee Glashow

Sheldon Lee Glashow is one among the many famous thinkers who have come through the Bronx High School of Science in New York, a testament to the capacity for vibrant public education institutions. Glashow is revered for his participation in two important breakthroughs in physics. First, along with Steven Weinberg and Abdus Salam, he developed a theory of electroweak interactions, which provides a unified theory for the previously distinct electromagnetic force and weak (nuclear) force. This was the work that garnered Glashow (as well as Weinberg—who was a classmate of his at Bronx Science and in undergraduate studies at Cornell University—and Salam) the 1979 Nobel Prize in Physics. In addition, this time in collaboration with John Iliopoulos and Luciano Maiani, Glashow predicted the charm quark, a variety of quark that, as a group, are one of the two basic constituents of matter. He is currently a professor at Boston University.

34. Lisa Randall

Lisa Randall is a leading theoretical physicist as well as a leader in breaking through glass ceilings in academia. In researching elementary particles and the fundamental forces of the universe, she has utilized a variety of theoretical models, including those that entail extra dimensions in space. She's applied her findings to make contributions to string theory, cosmological inflation, and general relativity, among other topics. Along the way she became the first woman to hold a tenured professorship in theoretical physics at, first, Princeton University, then the Massachusetts Institute of Technology, and now Harvard University, where she remains. In 2005 she published her first book, *Warped Passages: Unraveling the Mysteries of the Universe's Hidden Dimensions.*

35. Nicolás García

Nicolás García, a physicist, works at the CSIS (Center for Strategic and International Studies) in Madrid where he is expert in ballistic transport in nanostructures made of metal.

36. Paul Davies

Paul Davies is a theoretical physicist, cosmologist, and astrobiologist, and currently holds the position of Professor of Natural Philosophy in the Australian Centre for Astrobiology at Macquarie University. In 1995 he was awarded the Templeton Prize for his work on science and religion. He has written numerous books including *The Mind of God*, *The Accidental Universe*, and *How to Build a Time Machine*.

THE EDITORS

Eduardo Punset

After Eduardo Punset graduated from North Hollywood High School in Los Angeles in 1954, he received a law degree from Madrid University (La Complutense) in 1958. This was followed by a master's of science in economics at the London School of Economics in 1965.

For ten years he worked first in television for the BBC and later for the *Economist* in London. He worked in the International Monetary Fund (IMF) in Washington from 1969 to 1974 and was appointed their Caribbean representative. At the termination of the Generalissimo Franco regime he participated in the Spanish political transition to democracy and in the negotiations for Spanish entry to the European Union as Minister of European Relations. During the transition to democracy in Eastern Europe he was a member of the European Parliament for eight years and head of the Poland Committee of the European Parliament. From 1983 to 1990 he was chairman of a French multinational organization, the Bull Technological Institute. This pioneer institution assessed the impact of new technologies on economic management.

Appointed as professor of innovation and technology he taught at several business schools in Europe including the International Summer School at El Escorial, Spain, the former royal palace (Charles V, Philip II) outside Madrid, and the business institute (Instituto de Empresa) in Madrid. He has participated in many programs on science, technology, and society in Barcelona. Eduardo P. Punset is best known as a director and producer of *Redes* ("Networks"). He has hosted this popular weekly television program that communicates science to a worldwide Spanish-speaking audience since 1996. He currently also teaches as professor of science, technology, and society at the Ramon Llull University's Chemical Institute in Barcelona. As chairman of the board of Smart Planet he heads a multimedia science production company. He recently has been appointed president of the directorate on scientific communication of the Catalan Research Foundation, a granting agency of the provincial government in Barcelona. The aim of the agency is to advise and promote science communication of all kinds and at all levels by uniting scientific researchers, university professors, and media experts.

His 2004 book, *Cara a Cara con la vida, la mente y el Universo*, based on his worldwide travels and interviews with world-class scientists, and the basis of this book, is one of the best-selling science books in the Spanish language. Punset's *El viaje a la felicidad* (Barcelona: Ediciones Destino) has been translated as *The Happiness Trip* (Chelsea Green Publishing) and has sold more than 450,000 copies.

Punset was born in 1936 in Barcelona, Spain. He has three daughters, one of whom, Elsa, profoundly fluent in English as well as Spanish, has worked with him for over ten years. He has five grandchildren.

Lynn Margulis

Lynn Margulis, distinguished university professor in the Department of Geosciences at the University of Massachusetts–Amherst, received the 1999 National Medal of Science from President Bill Clinton. She has been a member of the U.S. National Academy of Sciences since 1983 and of the Russian Academy of Natural Sciences since 1997. Author, editor, or coauthor of chapters in more than forty books, she has published or been profiled in many journals, magazines, and books, among them *Natural History, Science, Nature, New England Watershed, Scientific American, Proceedings of the National Academy of Sciences, Science Firsts,* and *The Scientific 100*. She has made numerous contributions to the primary scientific literature of microbial evolution and cell biology.

Margulis's theory of species evolution by symbiogenesis, put forth in *Acquiring Genomes* (co-authored with Dorion Sagan, 2002), describes how speciation does not occur by random mutation alone but rather by symbiotic détente. Behavioral, chemical, and other interactions often lead to integration among organisms, members of different taxa. In well-documented cases some mergers create new species. Intimacy, physical contact of strangers, becomes part of the engine of life's evolution that accelerates the process of change. Margulis works in the laboratory and field with many other scientists and students to show how specific ancient partnerships, in a given order over millions of years, generated the cells of the species we see with our unaided eyes. The fossil record, in fact, does not show Darwin's predicted gradual changes between closely related species but rather the "punctuated equilibrium" pattern described by Eldredge and Gould: a jump from one to a different species.

She has worked on the "revolution in evolution" since she was a graduate student. Over the past fifteen years, Margulis has cowritten several books with Dorion Sagan, among them *What is Sex?* (1997), *What is Life?* (1995), *Mystery Dance: On the Evolution of Human Sexuality* (1991), *Microcosmos: Four Billion Years of Evolution from Our Microbial Ancestors* (1986), and *Origins of Sex: Three Billion Years of Genetic Recombination* (1986).

Her work with K.V. Schwartz and others provides a consistent formal classification of all life on Earth and has lead to the forthcoming fourth edition of *Five Kingdoms: An Illustrated Guide to the Phyla of Life on Earth*. Their classification scheme was generated from scientific results of myriad colleagues and its logical-genealogical basis is summarized in her single-authored book *Symbiosis in Cell Evolution: Microbial Communities in the Archean and Proterozoic Eons* (second edition, 1993). The bacterial origins of both chloroplasts and mitochondria are now well established. Currently, with colleagues and students, she explores the possible origin of cilia from spirochete bacteria.

Since the mid-1970s, Margulis has aided James E. Lovelock, FRS, in documenting his Gaia theory, which posits that the Earth's surface interactions among living beings, rocks, soil, air, and water have created a vast, self-regulating system. From the vantage of outer space the Earth looks like an amazing superorganism; from the vantage of biochemistry it behaves in many ways like a single giant ecosystem.

David T. Suzuki*

David T. Suzuki, a Canadian citizen, received his undergraduate degree in biology at Amherst College in 1958. He went on to earn his Ph.D. in zoology from the University of Chicago in 1961. He is currently founder and chair of the David Suzuki Foundation. As an award-winning scientist, he is best known in the scientific community as the author of a widely used textbook on genetics. He is a staunch environmentalist and an interviewer and producer of scientific films.

Dr. Suzuki receives consistent high acclaim for his thirty years of award-winning work in broadcasting. He can explain the complexities of science in a compelling, clear, and fascinating fashon. He is well known to millions in Canada and deserves to be known in the United States, Asia, and Europe far more than he is (because his programs are too scientific and cosmopolitan for most parochial and trivial television). He hosted *The Nature of Things*, a Canadian Broadcasting Corporation's science television series for twenty years. During his years as director of these science television programs in Canada he traveled all over the world, interviewed many scientists, he lived in the Amazon rain forest and climbed the Burgess Pass to see the splendid

*Suzuki is the author of the this book's Foreword but was not an editor of this book.

Cambrian fossils of the Shol. He studied the stromatolites of the marine Hamelin Pool in Western Australia. During most of the making of *The Nature of Things* he lived in Toronto. Most recently in his *Sacred Nature* series he has interviewed James Lovelock. His interviews have included Stephen J. Gould, and Lynn Margulis, whose work appears here.

His eight-part series *A Planet for the Taking* won an award from the United Nations. His eight-part PBS series *The Secret of Life* was praised internationally, as was his five-part series *The Brain* for the Discovery Channel. For CBC Radio he founded the long-running radio series *Quirks and Quarks* and has presented two influential documentary series on the environment, *From Naked Ape to Superspecies* and *It's a Matter of Survival.*

An internationally respected geneticist, David has been a full professor at the University of British Columbia (UBC) in Vancouver since 1969. When he is not teaching the rest of us through his public works he teaches his children and grandchildren. He is professor emeritus with UBC's Sustainable Development Research Institute; from 1969 to 1972 he was the recipient of the prestigious E.O.W.R. Steacie Memorial Fellowship Award for the Outstanding Canadian Research Scientist Under the Age of Thirty-five.

He has received numerous awards including the Roger Tory Peterson Award from Harvard University. He is a Companion of the Order of Canada, and a member of the Order of British Columbia. He has received eighteen honorary doctorates—twelve from Canada, four from the United States, and two from Australia. First Nations people have honored him with six names, formal adoption by two tribes, and made him an honorary member of the Dehcho First Nations.

Suzuki was born in Vancouver in 1936. During World War II, at the age of six, he was interned with his family in a camp in the British Columbia countryside for being of Japanese descent. After the war, he went to high school in London, Ontario. The author of forty-three books, David Suzuki is recognized as a world leader in sustainable ecology. He lives with his wife, Dr. Tara Cullis, and two daughters in Vancouver.

Dorion Sagan

Dorion Sagan, partner of Sciencewriters, is the author of numerous articles and sixteen books translated into eleven languages, including *Into the Cool: Energy Flow, Thermodynamics, and Life* (with Eric D. Schneider, 2005) and *Up from Dragons: Evolution of Human Intelligence* (with John Skoyles, 2002). His *What Is Life?* (with Lynn Margulis) was chosen (with works by Billie Holiday, Shakespeare, and others) as one of five "mind-altering masterpieces" by the *Utne Reader*. Sagan's essays are included in collections edited by Richard Dawkins and E. O. Wilson. He graduated from the University of Massachusetts–Amherst with a degree in history and has interests in philosophy and literature.

Reviewing Sagan's *Microcosmos* in the *New York Times Book Review*, Melvin Konner wrote: "This admiring reader of Carl Sagan, Lewis Thomas, and Stephen Jay Gould has seldom, if ever, seen such a luminous prose style in a work of this kind." Sagan has written for the *New York Times*, the *New York Times Book Review*, *Wired*, *Skeptical Inquirer*, *Pabular*, *Smithsonian*, *The Ecologist*, *Omni*, *Natural History*, and many others.

THE READINGS

We asked the interviewees to recommend further readings on the topics they discuss. Their suggestions are indicated in parentheses by their initials and the number of their interview. Readings without parentheses are from elsewhere in the text (e,g, David Suzuki's foreword, Eduardo Punset's comments, or the introductions to one of the four parts.)

Adams, Ansel, James Alinder, and John Szakowski. 1987. *Ansel Adams: Classic Images by Ansel Adams.* New York: Little Brown.

Asimov, Isaac. 1991. *Atom: Journey Across the Subatomic Cosmos.* New York: Truman Talley Books. (NG, 35)

Babiak, Paul, and Robert D. Hare. 2006. *Snakes in Suits: When Psychopaths Go to Work.* New York: HarperCollins. (RHa, 8)

Bonner, John Tyler. 1983. *Evolution of Culture in Animals.* Princeton, NJ: Princeton University Press. (JB, 30)

Brand, Stewart, and Peter Warshall. 1968–1998. *The Original Whole Earth Catalogue, Special 30th Anniversary Paperback.* Sausalito, CA: Point Foundation.

Brenner, Sydney. *My Life in Science.* 2001. London: BioMed Central.

Clark, R.W. 1981. *The Survival of Charles Darwin: A Biography of a Man and an Idea.* New York: Random House. (SJG, 25) (RHa, 8) (NM, 1) (RL, 11)

Chudnovsky, Eugene, and Javier Tejada. 2006. *Lectures on Magnetism.* Paramus, NJ: Rinton Press. (EC, 31)

Cole, K.C. 1999. *The Universe and the Teacup: The Mathematics of Truth and Beauty.* Fort Washington, PA: Harvest Books. (EC, 31)

Davies, Paul. 1995. *About Time.* London: Viking. (PD, 36)

Davies, Paul. 2001. *How to Build a Time Machine.* London: Allen Lane. (PD, 36)

Darwin, Charles. 1859. *On the Origin of Species by Means of Natural Selection.* London: J. Murray.

Dawkins, Richard. 2004. *The Ancestor's Tale.* Boston: Houghton Mifflin. (RD, 26)

Dawkins, Richard. 1999. *The Extended Phenotype: The Long Reach of the Gene.* New York: Oxford University Press. (RD, 26)

Day, William. 2002. *How Life Began.* Cambridge, MA: Foundation For New Directions. (WD, 28)

Day, William. Forthcoming 2008. "Life as Growth by Energy Flow." In Lynn Margulis and Celeste Asikainen, eds., *Chimeras and Consciousness: Evolution of the Sensory Self.* White River Junction, VT: Chelsea Green Publishing. (WD, 28)

de la Macorra, Ximena, and Antonio Vizciano, eds. 2006. *Water.* Mexico City: American Natural.

Dennett, Daniel. 1996. *Kinds of Minds.* New York: Basic Books. (DaD, 9)

Dennett, Daniel. 2005. *Sweet Dreams: Philosophical Obstacles to a Science of Consciousness.* Cambridge, MA: MIT Press. (DaD, 9)

Deutsch, Diana. *Musical Illusions and Paradoxes.* La Jolla, CA: Philomel Records, 1995. (DiD,17)

Deutsch, Diana. *Phantom Words and Other Curiosities*. La Jolla, CA: Philomel Records, 2003. (DiD, 17)

Deutsch, Diana. (Ed.) 1999. *The Psychology of Music*. San Diego, CA: Academic Press. (DiD,17)

Doyle, Gerald A., Jacopo Moggi-Cecchi, Michael A. Raath, and Philip V. Tobias, eds. 1999. *Humanity from Naissance to Coming Millenia*. Johannesburg: Witswatersrand University Press. (PT, 20)

Fleck, Ludwik. 1979. *Genesis and Development of a Scientific Fact*. Chicago: University of Chicago Press. (JB, 30)

Gest, Howard. 2003. *Microbes: An Invisible Universe*. Washington, DC: ASM Press. (RGu, 29)

Gilbert, Daniel. 2006. *Stumbling on Happiness*. New York: Knopf. (DG, 7)

Glashow, Sheldon Lee. 1991. *The Charm of Physics*. Melville, NY: AIP Press. (SLG, 33)

Glashow, Sheldon Lee. 1994. *From Alchemy to Quarks*. Belmont, CA: Brooks/Cole Publishing Company. (SLG, 33)

Goleman, Daniel. 1997. *Emotional Intelligence: Why It Can Matter More Than IQ*. New York: Bantam. (RL, 11)

Goodall, Jane. 1988. *In the Shadow of Man*. New York: Houghton Mifflin. (JG, 3)

Gould, Stephen Jay. 1989. *The Panda's Thumb: More Reflections in Natural History*. New York: W. W. Norton & Company. (SJG, 25)

Gould, Stephen Jay. 1980. *Wonderful Life: The Burgess Shale and the Nature of History*. New York: W. W. Norton & Company. (SJG, 25)

Gregory, Richard L. 1997. *Eye and Brain: The Psychology of Seeing*. Oxford: Oxford University Press. (RGr, 15)

Haraway, Donna. 1989. *Primate Visions: Gender, Race, and Nature in the World of Modern Science*. New York: Routledge. (JG, 3)

Harding, Stephan. 2006. *Animate Earth: Science, Intuition, and Gaia*. White River Junction, VT: Chelsea Green Publishing.

Hare, Robert D. 1998. *Without Conscience: The Disturbing World of the Psychopaths Among Us*. New York: Guilford. (RHa, 8)

Holloway, Ralph L. Jr. 1992. "Culture: A Human Domain." *Current Anthropology*, Vol. 33, No. 1, pp. 47-64 (*Supplement: Inquiry and Debate in the Human Sciences: Contributions from Current Anthropology, 1960–1990*). (RHo, 20)

Holloway, R.L., Yuan, M.S., and Broadfield, D.C. *Brain Endocasts: Paleoneurological Evidence*. A Volume in Jeff Schwartz and Ian Tattersall's *Human Fossil Record* series. New York: John Wiley & Sons. (RHo, 20)

The Human Genome. 2001. A special issue of *Science* magazine devoted to the completion of the human genome map. Vol. 291, Issue 5507 (February 16, 2001). www.sciencemag.org/content/vol291/issue5507/ (WH, 19)

Humphrey, Nicholas. 2002. "A Book at Bedtime." In Nicholas Humphrey, *The Inner Eye*. Oxford: Oxford University Press. (NH, 16)

Johnston, Victor S. 2006. "Mate choice decisions: the role of facial beauty." *Trends in Cognitive Science*. Vol. 10, No. 1, pp. 9-13. (VJ, 6)

Johnston, Victor S. 1999. *Why We Feel: The Science of Human Emotions*. New York: Perseus Press. (VJ, 6)

Kendler, Kenneth S. 2006. *Genes, Environment, and Psychopathology: Understanding the Causes of Psychiatric and Substance Use Disorders*. New York: Guilford Press. (KK, 13)

Kendrick, Bryce. 2001. *The Fifth Kingdom.* Newburyport, MA: Focus Publishing/R. Pullins Company. (KK, 13)

Kirkwood, Tom. 2003. *Time of Our Lives: The Science of Human Aging.* New York: Oxford University Press. (TK, 22)

Kirkwood, Tom. 2005. "Time of our lives: What controls the length of life?" *EMBO Reports.* Vol. 6, *Special Issue: Science & Society*, pp. S4–S8. www.nature.com/embor/journal/v6/n1s/full/7400419.html (TK, 22)

Konner, Melvin. 2003. *The Tangled Wing: Biological Constraints on the Human Spirit.* New York: Owl Books. (RS, 2)

Kuhn, Thomas. 1985. *The Structure of Scientific Revolutions.* Chicago: University of Chicago Press. (JB, 30)

Lapo, André V. 1987. *Traces of Bygone Biospheres.* Moscow: Mir.

LeDoux, Joseph. 1996. *The Emotional Brain.* New York: Simon and Schuster. (JL, 12)

LeDoux, Joseph. 2002. *Synaptic Self: How Our Brains Become Who We Are.* New York: Viking. (JL, 12)

Llinás, Rodolfo. 2002. *I of the Vortex: From Neurons to Self.* Cambridge, MA: MIT Press. (RL, 11)

Llinás, Rodolfo. 1999. *The Squid Giant Synapse: A Model for Chemical Transmission.* New York: Oxford University Press. (RL, 11)

Llinás, Rodolfo, and Mircea Steriade. 2006. "Bursting of Thalamic Neurons and States of Vigilance." *Journal of Neurophysiology.* Vol. 95, No. 6, pp. 3297-3308. http://jn.physiology.org/cgi/content/abstract/95/6/3297 (RL, 11)

Lovelock, James. 1995. *The Ages of Gaia: A Biography of Our Living Earth.* New York: W. W. Norton & Company. (JEL, 23)

Lovelock, James. 2001. *Homage to Gaia.* New York: Oxford University Press. (JEL, 23)

Lovelock, James. 2007. *The Revenge of Gaia: Earth's Climate Crisis and the Fate of Humanity.* New York: Basic Books. (JEL, 23)

Mackintosh, Nicholas. 1996. "Intelligence in Evolution." In Jean Khalfa (ed.), *What Is Intelligence?* New York: Cambridge University Press. (NM, 1)

Mackintosh, Nicholas. 1998. *IQ and Human Intelligence.* New York: Oxford University Press. (NM, 1)

Margulis, Lynn. 1993. *Symbiosis in Cell Evolution*, 2nd ed. New York: W.H. Freeman & Company. (DW, 21)

Margulis, Lynn. 2000. *Symbiotic Planet.* New York: Basic Books.

Margulis, Lynn, and Celeste Asikainen, eds. Forthcoming 2008. "Bacteria and Protists Also Make Decisions." In *Chimeras and Consciousness: Evolution of the Sensory Self.* White River Junction, VT: Chelsea Green Publishing. (NG, 35)

Margulis, Lynn, and Dorion Sagan. 2002. *Acquiring Genomes: A Theory of the Origins of Species.* New York: Basic Books. (EC, 31)

Margulis, Lynn, and Dorion Sagan. 2007. *Dazzle Gradually: Reflections on the Nature of Nature.* White River Junction, VT: Chelsea Green Publishing. (SLG, 33)

Margulis, Lynn, and James E. Lovelock. 1975. "The atmosphere as circulatory system of the biosphere: the Gaia hypothesis." *CoEvolution Quarterly*, Summer 1975, pp. 30–40. Reprinted in Lynn Margulis and Dorion Sagan, *Dazzle Gradually: Reflections on the Nature of Nature.*

Margulis, Lynn, and M.J. Chapman. Forthcoming, 2008. *Kingdoms and Domains: An Illustrated Guide to the Phyla of Life on Earth,* 4th ed. San Diego: Academic Press–Elsevier.

Miller, Jonathan. 1986. *The Body in Question.* New York: Horizon Book Promotions. (RL, 11)

Nealson, Kenneth H. 2001. "Searching for life in the universe: Lessons from the Earth." *Cosmic Questions: Annals of the New York Academy of Sciences.* Vol. 950, pp. 241–258. www.annalsnyas.org/cgi/content/abstract/950/1/241 (KN, 24)

Nealson, Kenneth H., and Pamela G. Conrad. 1999. "Life: past, present and future." *Philosophical Transactions of the Royal Society B: Biological Sciences.* Vol. 354, No. 1392, pp. 1923–1939. www.journals.royalsoc.ac.uk/link.asp?id=7r10hqn3rp1g1vag (KN, 24)

Punset, Eduardo. 2005. *El viaje a la felicidad.* Barcelona: Ediciones Destino. In translation (2007) as *The Happiness Trip.* White River Junction, VT: Chelsea Green Publishing. (RS, 2)

Randall, Lisa. 2005. *Warped Passages: Unraveling the Mysteries of the Universe's Hidden Dimension.* New York: Ecco. (LR, 34)

Sabater Pi, Jordi. 1992. *El chimpancé y los orígenes de la cultura.* Barcelona: Anthropos. (JSP, 4)

Sabater Pi, Jordi. 1993. *Gorilas y chimpancés del África Occidental.* Mexico City: Fondo de cultura económica. (JSP, 4)

Sacks, Oliver. 2002. *Oaxaca Journal.* New York: National Geographic Directions. (OS, 10)

Sacks, Oliver. 2001. *Uncle Tungsten.* New York: Knopf. (OS, 10)

Sagan, Carl. 1996. *The Demon-Haunted World: Science as a Candle in the Dark.* New York: Random House. (SLG, 33)

Sagan, Dorion, and Lynn Margulis. 2006. "Water and Life." In Ximena de la Macorra and Antonio Vizcaino, eds., *Water.* Mexico City: American Natural.

Sagan, Dorion, and Jessica Hope Whiteside. 2005. "Gradient reduction theory: Thermodynamics and the purpose of life." In Stephen H. Schneider, James R. Miller, Eileen Crist, and Penelope J. Boston, eds., *Scientists Debate Gaia: The Next Century.* Cambridge, MA: MIT Press. (DS, 27) (SB, 18)

Sapolsky, Robert. 2004. *Why Zebras Don't Get Ulcers: The Acclaimed Guide to Stress, Stress-Related Diseases, and Coping.* New York: Owl Books. (RS, 2)

Schaechter, Moselio, John L. Ingraham, and Frederick C. Neidhardt. 2006. *Microbe.* Washington, DC: ASM Press. (RGu, 29)

Schneider, Eric D., and Dorion Sagan. 2005. *Into the Cool: Energy Flow, Thermodynamics, and Life.* Chicago: University of Chicago Press. (DS, 27) (SB, 18) (EC, 31)

Scientific American, the editors of. 2002. *Understanding Nanotechnology.* New York: Warner Books. (HR, 32)

Sober, Elliot, and David Sloan Wilson. 2007. *Unto Others: The Evolution and Psychology of Unselfish Behavior.* Cambridge, MA: Harvard University Press.

Stewart, Melissa. 1999. *Life without Light: A Journey to Earth's Dark Ecosystems* London: Franklin Watts. (EOW, 5)

Strogatz, Steven. 2003. *SYNC: The Emerging Science of Spontaneous Order.* New York: Hyperion. (SS, 14)

Thorne, Kip. 1995. *Black Holes and Time Warps: Einstein's Outrageous Legacy.* New York: W. W. Norton & Company. (LR, 34)

U.S. Department of Energy Office of Science Genome Programs. 2003. "The Human Genome and Beyond." http://web.ornl.gov/sci/techresources/Human_Genome/publicat/primer2pager.pdf. This two-page flyer gives an overview of the Human Genome Project and lists sources for more information at different levels of complexity and difficulty. (WH, 19)

Vernadsky, Vladimir I. 1998. *The Biosphere*. (Translated from Russian; original edition, 1926.) New York: Simon & Schuster.

von Baeyer, Hans Christian. 2000. *Taming the Atom: The Emergence of the Visible Microworld*. Mineola, NY: Dover. (HR, 32)

Wallace, Douglas. 1997. "Mitochondrial DNA in Aging and Disease." *Scientific American*. Vol. 227, No. 2, pp. 40–47. (DW, 21)

Westbroek, Peter. 1992. *Life as a Geological Force: Dynamics of the Earth*. New York: W. W. Norton & Company.

Wilson, Edward O. 1998. *Consilience: The Unity of Knowledge*. New York: Knopf. (EOW, 5)

Wilson, Edward O. 2000. *Sociobiology: The New Synthesis: Twenty-fifth Anniversary Edition*. Cambridge, MA: Belknap/Harvard University Press (EOW, 5)

Wood, William B. 1988. *The Nematode* Caenorhabditis elegans. Woodbury, NY: Cold Spring Harbor Laboratory Press. (SB, 18)

LIST OF FIGURES AND TABLES

INDEX